GEOMETRY OF RIEMANN SURFACES
AND TEICHMÜLLER SPACES

NORTH-HOLLAND MATHEMATICS STUDIES 169

(Continuation of the Notas de Matemática)

NORTH-HOLLAND – AMSTERDAM • LONDON • NEW YORK • TOKYO

GEOMETRY OF RIEMANN SURFACES AND TEICHMÜLLER SPACES

Mika SEPPÄLÄ

Academy of Finland
Helsinki, Finland

Tuomas SORVALI

University of Joensuu
Joensuu, Finland

1992

NORTH-HOLLAND – AMSTERDAM • LONDON • NEW YORK • TOKYO

ELSEVIER SCIENCE PUBLISHERS B.V.
Sara Burgerhartstraat 25
P.O. Box 211, 1000 AE Amsterdam, The Netherlands

Distributors for the United States and Canada:

ELSEVIER SCIENCE PUBLISHING COMPANY, INC.
655 Avenue of the Americas
New York, N.Y. 10010, U.S.A.

Library of Congress Cataloging-in-Publication Data

Seppälä, Mika.
Geometry of Riemann surfaces and Teichmüller spaces / Mika Seppälä, Tuomas Sorvali.
p. cm. -- (North-Holland Mathematics studies ; 169)
Incluoes bibliographical references and index.
ISBN 0-444-88846-2
1. Riemann surfaces. 2. Teichmüller spaces. I. Sorvali, Tuomas, 1944- . II. Title. III. Series.
QA333.S42 1992
515'.223--dc20 91-34760
CIP

ISBN: 0 444 88846 2

Printed in The Netherlands

Preface

This monograph grew out of a series of lectures held by the first author at the University of Regensburg in 1986 and in 1987 and by the second author at the University of Joensuu in 1990. This book would presumably not have been written without the initiative of Professor Leopoldo Nachbin.

A large part of the present work has been carried out at the University of Regensburg and at the Mittag–Leffler Institute. We thank these both institutes for their warm hospitality.

Finally we thank Ari Lehtonen for several figures, especially for his intriguing illustration of the Klein bottle.

In Helsinki and in Joensuu, Finland

August 1991

Mika Seppälä Tuomas Sorvali

Introduction

The moduli problem is to describe the structure of the space of isomorphism classes of Riemann surfaces of a given topological type. This space is known as the *moduli space*. It has been in the center of pure mathematics for more than 100 years now. In spite of its age, this field still attracts lots of attention. The reason lies in the fact that smooth compact Riemann surfaces are simply complex projective algebraic curves. Therefore the moduli space of compact Riemann surfaces is also the moduli space of complex algebraic curves. This space lies in the intersection of many fields of mathematics and can, therefore, be studied from many different points of view.

Our aim is to get information about the structure of the moduli space using as concrete and as elementary methods as possible. This monograph has been written in the classical spirit of Fricke and Klein ([31]) and in that of Lehner ([57]). Our main goal is to see how far the concrete computations based on uniformization take us. It turns out that this simple approach leads to a rich theory and opens a new way of treating the moduli problem. Or rather puts new life in the classical methods that were used in the study of moduli problems already in the 1920's.

Some results, like the Uniformization of Riemann surfaces, have to be presented here without proofs. They are, however, used almost exclusively to interpret the results derived by other means. Proofs are not really based on them. In all cases, where we do not present proofs, we furnish exact references.

If one is willing to accept Uniformization and some related facts, then this monograph is self–contained and can be read without much prior knowledge about complex analysis.

In Chapter 1 we develop an engine that will power other chapters. There we consider Möbius transformations and matrices. One of our aims in Chapter 1 is to understand thoroughly how commutators of Möbius transformations behave and how groups generated by Möbius transformations can be parametrized. All considerations here are elementary, but sometimes technically complicated.

In Chapter 2 we present some basic results of the theory of quasiconformal mappings. Everything there is presented without proofs, which can be found in the monograph of Lars V. Ahlfors [6] and in that of Olli Lehto and Kalle Virtanen [59]. Quasiconformal mappings have played an important role in the theory of Teichmüller spaces. They provided the tools with which it was possible to develop the first rigorous treatment of the moduli problem. Today most of the results concerning Teichmüller spaces and moduli spaces can be shown even without quasiconformal mappings. Quasiconformal mappings are only absolutely necessary to show that the moduli space of symmetric Riemann surfaces of a given topological type is connected (cf. Theorem 4.4.1 on page 147).

In Chapter 3 we first review the Uniformization of Riemann surfaces without proofs. Then we show how considerations of Chapter 1 can be applied to study the geometry of Riemann surfaces. Our main concern in this Chapter is to study the geometry of hyperbolic metrics of Riemann surfaces of negative Euler characteristics. We derive many results concerning simple closed geodesics and sizes of collars around them. We pay special attention to the geometry of symmetric Riemann surfaces, i.e., to non–classical Klein surfaces.

Everything here can be shown in detail using the results of Chapter 1. It is actually surprising how much information can be obtained from detailed analysis of the commutator of Möbius transformations. Considerations of Chapter 3 form a quite comprehensive treatment of certain aspects of the geometry of hyperbolic surfaces. So it may be of some interest for its own sake already. Main target is, however, to get information about the moduli problem using considerations of Chapter 1 alone. The beginning of Chapter 3 provides an environment in which considerations of Chapter 1 can be interpreted so that we get useful results for later applications.

In Chapter 4 we introduce Teichmüller spaces and define its topology using quasiconformal mappings. Here we have to resort to the review presented without proofs in Chapter 2. We will, however, later derive an alternative way of parametrizing the Teichmüller space using the geodesic length functions. That is done in detail here (see page 161).

Quasiconformal mappings provide a simple way to describe the complex structure of the Teichmüller space of classical Riemann surfaces (cf. page 148). We will take benefit of that description and indicate how our considerations lead to a *real analytic theory* of Teichmüller spaces. This also leads to a presentation of Teichmüller spaces as a component of an affine real algebraic variety (Section 4.12). This affine structure is derived here in detail. This is an important part in the theory of Teichmüller spaces, albeit not central, because it opens new ways of compactifying the Teichmüller space by using methods of real algebraic geometry (cf. [64], [16], [73]). We will not consider these interesting approaches to the compactification problem here.

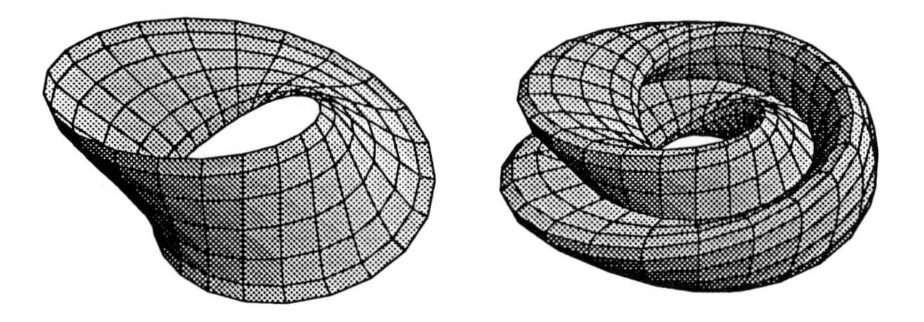

Figure 0.1: The Möbius strip and the Klein bottle are two genus 1 real algebraic curves that are not homeomorphic to each other. In this monograph a new moduli space is constructed for the these non–classical Klein surfaces.

The presentation of the affine structure of Teichmüller spaces is, however, partly motivated by these new applications of real algebraic geometry.

Points of the moduli space of compact genus g Riemann surfaces are isomorphism classes of mutually homeomorphic genus g Riemann surfaces. Such Riemann surfaces are smooth projective complex algebraic curves. So the moduli space of genus g Riemann surfaces is the same thing as the moduli space of smooth genus g complex algebraic curves.

In Chapter 5 we consider this moduli space and define a natural topology for it. The definition of the topology is based on the *Fenchel–Nielsen coordinates.* In that topology the moduli space is connected but not compact. Using the considerations of Chapter 3 we then consider degenerating sequences of Riemann surfaces. It turns out that by adding points corresponding to so called stable Riemann surfaces it is possible to compactify the moduli space of compact and smooth genus g Riemann surfaces. This is quite classical today and has first been shown by David Mumford and others using the methods of complex algebraic geometry.

Smooth projective *real* algebraic curves have more structure than complex curves. They can be viewed as compact Riemann surfaces *with symmetry.* Equally well they can be viewed as compact *non–classical Klein surfaces*, i.e., surfaces that are obtained as the quotient of a smooth Riemann surface by the action of the symmetry. This fact was realized already by Felix Klein (cf. [46]). Therefore the moduli spaces of *non–classical* compact Riemann surfaces are simply moduli spaces of real algebraic curves.

A compact genus g surfaces has $\lfloor(3g+4)/2\rfloor$ topologically different orientation reversing symmetries. It follows, especially, that real algebraic curves of the same genus need not be homeomorphic to each other. This implies that, in any reasonable topology, the moduli space of smooth genus g real algebraic curves has several connected components.

The situation changes completely when we consider the natural compactification of the moduli space of real algebraic curves. That space is obtained by adding points corresponding to stable genus g real algebraic curves. We show, in Chapter 5, that this moduli space of stable real algebraic curves of a given genus g, $g > 1$, is a connected and compact Hausdorff space. This fact was already conjectured by Felix Klein in [48, Page 8].

We start Chapter 1 with the assumption that the reader is familiar with elementary properties of Möbius transformations. For the sake of completeness we have included also an appendix in which we develop the elementary and classical theory of Möbius transformations. A proof for the so called Nielsen Criterium for discreteness of Möbius groups acting in a disk is also included. Basic properties of the hyperbolic geometry are considered in an appendix as well.

Resorting to the appendices, if necessary, this monograph can be read with only basic knowledge of complex analysis.

Contents

Chapter 1

Geometry of Möbius transformations

1.1 Introduction to Chapter 1

Our main interest lies in parametrizing groups generated by Möbius transformations. By the Uniformization such a parametrization can be used to study surfaces and their complex structures. That is the theme of chapters 3, 4 and 5. In this chapter we derive the necessary preliminary results which then power the rest of the book (excluding Chapter 2).

First we recall briefly the classification of Möbius–transformations. That is explained in more detail in Appendix B in which we review the classical theory of Möbius transformations and groups of Möbius transformations.

Natural question is to find ways of parametrizing Möbius groups *up to a conjugation by a Möbius transformation.* Therefore we need parameters which remain invariant under such conjugations. Natural candidates for such parameters are the *multipliers* of Möbius transformations. Equivalently one may use also *traces* of the corresponding matrices.

In this chapter we will mainly use multipliers. They are more natural since they are *uniquely* determined while the trace of the corresponding matrix is determined only up to the *sign* by the Möbius transformation. This difference may at first sight appear only as a small technical complication, but it turns out to be one of the main difficulties. The problem lies in the fact that the sign the trace of a product of two matrices is not determined by the signs of traces of the matrices in question. To deal with that difficulty we introduce, in Section 1.4, a classification of pairs of hyperbolic Möbius transformations. We use that classification to find natural parameters which determine a group (generated by a finite number of hyperbolic Möbius transformations) up to a conjugation by a Möbius transformation.

The main problem of this section is to find a *minimal* set of parameters for Möbius groups generated by a finite number of hyperbolic Möbius transformations. That is also important for later applications. In the general case these groups are not freely generated. The generators usually satisfy a relation which typically says that the product of certain commutators is the identity. Such a relation is difficult to deal with. In the general case the problem of finding a *minimal* parametrization using only multipliers of elements of the group remains open. In Theorem 1.6.4 we summarize our results concerning this problem. It is our conjecture that the parametrization given by Theorem 1.6.4 is actually minimal. Recent investigations by Chen Min support this conjecture, but it has not been shown yet.

A minimal parametrization can be constructed if we use also certain *twist* parameters or *gluing angles.* That is done in Section 1.7. The twist parameters presented there are closely related to the usual *Fenchel–Nielsen gluing angles*[1], but they are not the same.

1.2 Möbius transformations

We shall consider groups of conformal automorphisms of the extended complex plane $\hat{\mathbf{C}}$. Directly conformal automorphisms of $\hat{\mathbf{C}}$ are *orientation preserving Möbius transformations*

$$z \mapsto \frac{az+b}{cz+d}, \quad ad-bc=1, \tag{1.1}$$

whereas indirectly conformal ones are *orientation reversing Möbius transformations*

$$z \mapsto \frac{a\bar{z}+b}{c\bar{z}+d}, \quad ad-bc=-1. \tag{1.2}$$

We shall concentrate on groups which act in the upper half–plane U. It is a well–known exercise in complex analysis to show that transformations (1.1) and (1.2) fix U if and only if the coefficients a, b, c and d are real. Rays or half–circles in U perpendicular to the real axis $\mathbf{R}$ are called *non–Euclidean lines.* A Möbius transformation fixing U maps a non–Euclidean line onto a non–Euclidean line.

There are the following *types* of Möbius transformations fixing U:

- the identity transformation,
- hyperbolic transformations,
- parabolic transformations,

[1] For the usual definition of the Fenchel–Nielsen gluing angles see Section 5.5

- elliptic transformations,
- reflections,
- glide–reflections.

The first four types are orientation preserving whereas the last two ones are orientation reversing.

It will turn out that *hyperbolic* transformations are the most essential ones. Geometrically, the hyperbolic transformation

$$g(z) = \frac{az+b}{cz+d}, \quad ad - bc = 1, \tag{1.3}$$

is determined by the following three parameters:

- the *attracting* fixed point $a(g) = \lim_{n\to\infty} g^n(z)$
- the *repelling* fixed point $r(g) = \lim_{n\to\infty} g^{-n}(z)$
- the *multiplier*

$$k(g) = (g(z), z, r(g), a(g)) = \frac{g(z) - r(g)}{g(z) - a(g)} \cdot \frac{z - a(g)}{z - r(g)},$$

where z is any point in $\hat{\mathbf{C}}$ not fixed by g. Especially, $k(g) > 1$. The fixed points are real if and only if $g(U) = U$. In this case, the non–Euclidean line through $a(g)$ and $r(g)$, the *axis* of g, is denoted by $ax(g)$. It has natural orientation by $r(g) \to a(g)$.

Denoting $k = k(g)$, $x = a(g)$ and $y = r(g)$ we obtain, from the cross–ratio defining $k(g)$, the following representations for g:

$$g(z) = \frac{(kx-y)z - xy(k-1)}{(k-1)z + x - ky} \quad \text{if} \quad x \neq \infty \neq y, \tag{1.4}$$

$$g(z) = kz - y(k-1) \quad \text{if} \quad x = \infty, \tag{1.5}$$

$$g(z) = \frac{z}{k} + x(1 - \frac{1}{k}) \quad \text{if} \quad y = \infty. \tag{1.6}$$

Conversely, if $k > 0$, x and y, $x \neq y$, are given, then the Möbius transformation g defined by formulae (1.4) has the following properties:

- if $k = 1$, then $g = \mathrm{id}$,
- if $k > 1$, then g is hyperbolic, $k(g) = k$, $a(g) = x$ and $r(g) = y$,
- if $k < 1$, then g is hyperbolic, $k(g) = 1/k$, $a(g) = y$ and $r(g) = x$.

Since the coefficients in (1.3) are determined up to the sign, it follows by the formulae (1.4) that

$$|a+d| = \sqrt{k(g)} + \frac{1}{\sqrt{k(g)}} = \sqrt{k(g) + \frac{1}{k(g)} + 2}. \tag{1.7}$$

Hence $|a+d| > 2$.

Consider *conjugate* Möbius transformations g and

$$g' = h \circ g \circ h^{-1}$$

where $h : \hat{\mathbf{C}} \mapsto \hat{\mathbf{C}}$ is a Möbius transformation. Then g and g' are of the same type. Suppose that g is hyperbolic and fixes U. Then

- g' fixes $h(U)$,
- $a(g') = h(a(g))$ and $r(g') = h(r(g))$,
- $ax(g') = h(ax(g))$,
- h maps the non–Euclidean lines of U onto the non–Euclidean lines of $h(U)$.

The conjugacy class of a hyperbolic transformation is determined by its multiplier. In fact, for any hyperbolic transformations

$$g(z) = \frac{az+b}{cz+d}, \quad ad - bc = 1,$$

and

$$g'(z) = \frac{a'z+b'}{c'z+d}, \quad a'd' - b'c' = 1,$$

the following conditions are equivalent:

- g and g' are conjugate,
- $k(g) = k(g')$,
- $|a+d| = |a'+d'|$.

Let us consider the other types of Möbius transformations fixing U. A *parabolic* transformation has one fixed point only. For (1.1) this occurs if and only if $|a+d| = 2$. If we set $k(g) = 1$ in the parabolic case, then (1.7) remains valid.

Elliptic transformations are conjugate to rotations

$$z \mapsto e^{i\vartheta} z \tag{1.8}$$

of the complex plane. If g_1 and g_2 are conjugate to

$$z \mapsto e^{i\vartheta_1} z \quad \text{and} \quad z \mapsto e^{i\vartheta_2} z,$$

respectively, then g_1 and g_2 are conjugate if and only if $\vartheta_1 = \pm\vartheta_2 + n2\pi$. Therefore, we may define the multiplier $k(g)$ of an elliptic g conjugate to (1.8) by setting

$$k(g) = e^{i\vartheta}, \quad \text{where} \quad 0 < \vartheta \leq \pi.$$

If we denote by x and y the fixed points of g, then either

$$e^{i\vartheta} = (g(z), z, x, y)$$

or

$$e^{-i\vartheta} = (g(z), z, x, y)$$

for all $z \neq x$, y. Hence formulae (1.4) hold also in the elliptic case either with $k = e^{i\vartheta}$ or with $k = e^{-i\vartheta}$. Moreover, since

$$0 \leq e^{i\vartheta} + e^{-i\vartheta} + 2 < 4$$

for all $\vartheta \in \mathbf{R}$, it follows that (1.7) is valid and $|a + d| < 2$.

An elliptic transformation fixes U if and only if its fixed points are complex conjugates.

The composition of two orientation preserving transformations

$$g(z) = \frac{az + b}{cz + d}, \quad ad - bc \neq 0,$$

and

$$h(z) = \frac{\alpha z + \beta}{\gamma z + \delta}, \quad \alpha\delta - \beta\gamma \neq 0,$$

is obtained by multiplying the corresponding matrices:

$$\begin{pmatrix} a & b \\ c & d \end{pmatrix} \begin{pmatrix} \alpha & \beta \\ \gamma & \delta \end{pmatrix} = \begin{pmatrix} a\alpha + b\gamma & a\beta + b\delta \\ c\alpha + d\gamma & c\beta + d\delta \end{pmatrix}. \tag{1.9}$$

Hence

$$g(h(z)) = \frac{(a\alpha + b\gamma)z + a\beta + b\delta}{(c\alpha + d\gamma)z + c\gamma + d\delta}.$$

For orientation reversing transformations this is not true in general. But restricting ourselves to transformations with real coefficients, the coefficients of the composite transformation are obtained by (1.9) regardless of whether any of the transformations is orientation reversing.

The axis

$$ax(\sigma) = \{z \mid \sigma(z) = z\}$$

of a *reflection* σ fixing U is a circle or line perpendicular to $\mathbf{R}$, i.e., a non–Euclidean line in U. Denote by x and y the real fixed points of σ. Then

$$\sigma(z) = \psi(\overline{z})$$

where ψ is the elliptic transformation defined by $k(\psi) = -1$, $\psi(x) = x$ and $\psi(y) = y$. Hence, inserting $k = -1$ in formulae (1.4), we obtain the following representations:

$$\sigma(z) = \frac{(x+y)\overline{z} - 2xy}{2\overline{z} - (x+y)} \quad \text{if} \quad x \neq \infty \neq y \tag{1.10}$$

$$\sigma(z) = -\overline{z} + 2y \quad \text{if} \quad x = \infty \tag{1.11}$$

Especially, σ and ψ agree on the real axis.

A *glide–reflection* s fixing U is of the form

$$s = \psi \circ \sigma$$

where

- ψ is a hyperbolic transformation fixing U,
- σ is a reflection fixing U,
- $ax(\psi) = ax(\sigma)$.

Hence the glide–reflection s is uniquely determined by ψ. Moreover, since $s^2 = \psi^2$, also the hyperbolic transformation s^2 defines s uniquely. Especially, s and ψ have the same fixed points and the same axis.

Denote $x = a(s)$, $y = r(s)$ and $k = k(s) = -k(\psi)$. Formulae (1.4) and (1.10) then yield the following representations:

$$s(z) = \frac{(kx - y)\overline{z} - xy(k-1)}{(k-1)\overline{z} + x - ky} \quad \text{if} \quad x \neq \infty \neq y, \tag{1.12}$$

$$s(z) = k\overline{z} - y(k-1) \quad \text{if} \quad x = \infty, \tag{1.13}$$

$$s(z) = \frac{\overline{z}}{k} + x(1 - \frac{1}{k}) \quad \text{if} \quad y = \infty. \tag{1.14}$$

If we define $k(\sigma) = -1$ for a reflection σ, then formulae (1.10) are obtained from (1.12) and (1.13) as special cases. It follows that

$$|a+d| = \sqrt{|k(s)|} - \frac{1}{\sqrt{|k(s)|}}$$

for all transformations (1.2) fixing U. For reflections we have $|a+d| = 0$ whereas $|a+d|$ is positive for all glide–reflections.

A transformation (1.4) with $k < -1$ is *loxodromic.* A loxodromic transformation g has well-defined fixed points $a(g)$ and $r(g)$ and hence also a well-defined multiplier $k(g)$. By formulae (1.12), a glide-reflection s fixing U admits also a representation

$$s(z) = \psi(\overline{z})$$

where ψ is the loxodromic transformation defined by $k(\psi) = k(s)$, $a(\psi) = a(s)$ and $r(\psi) = r(s)$. On the real axis, s and ψ agree. Note that ψ maps U onto the lower half-plane and that ψ is decreasing on the real axis.

1.3 Multiplier preserving isomorphisms

In this section, we consider groups G of Möbius transformations acting in U. We are interested in developing conformally invariant systems of identification for the groups G. Our final goal will be to define a minimal set of identification numbers. In fact, in the next sections, we will give a "social security vector" to every group with a certain normalization (Theorem 1.5.6). For precise formulation of the results we consider isomorphisms of the groups G.

The next lemma shows that the Möbius group G is in most cases determined by its hyperbolic elements.

Lemma 1.3.1 *Let g and h be Möbius transformations fixing U, h hyperbolic and $g(a(h)) \neq r(h)$. Then $g \circ h^n$ is hyperbolic or a glide-reflection for sufficiently large values of n.*

Proof. We may suppose that $a(h) = \infty$ and $r(h) = 0$. Let $k = k(h)$ and consider the representation (1.1) or (1.2) of g. Then

$$g \circ h^n = \begin{pmatrix} a & b \\ c & d \end{pmatrix} \begin{pmatrix} \sqrt{k^n} & 0 \\ 0 & 1/\sqrt{k^n} \end{pmatrix} = \begin{pmatrix} a\sqrt{k^n} & b/\sqrt{k^n} \\ c\sqrt{k^n} & d/\sqrt{k^n} \end{pmatrix}.$$

Since $g(\infty) \neq 0$, we have $a \neq 0$. Hence

$$|a\sqrt{k^n} + \frac{d}{\sqrt{k^n}}| > 2 \quad \text{for} \quad n \geq n_0.$$

It follows that $g \circ h^n$ is either hyperbolic or a glide-reflection for $n \geq n_0$. □

If, in the above lemma, $g \circ h^n$ is a glide-reflection, then it is determined by the hyperbolic transformation $(g \circ h^n)^2$. Hence g is determined by the hyperbolic transformations h and $(g \circ h^n)^2$ for any $n \geq n_0$.

We show next that, under quite general assumptions, the group G is in fact determined up to conjugation already by the multipliers of its hyperbolic elements. To that end we need some technical lemmas.

The function

$$f(k) = \sqrt{k + \frac{1}{k} + 2} = \sqrt{k} + \frac{1}{\sqrt{k}}$$

is well-defined and non-negative for both $k > 0$ and $k = e^{i\vartheta}$. We have

- $f(k_1) = f(k_2) \Longleftrightarrow k_1 = k_2^{\pm 1}$,
- $f(k) \geq f(1) = 2 \quad \text{for} \quad k > 0$,
- $0 = f(-1) \leq f(k) \leq f(1) = 2 \quad \text{for} \quad k = e^{i\vartheta}$,
- $f(k) \to \infty \quad$ if and only if $\quad \max(k, 1/k) \to \infty$.

Hence

$$f(g) = f(k(g))$$

is defined for all orientation preserving transformations

$$g(z) = \frac{az + b}{cz + d}, \quad ad - bc = 1$$

fixing a disk or a half–plane and

- $f(g) = |a + d|$,
- $f(g) > 2 \Longleftrightarrow g$ hyperbolic,
- $f(g) = 2 \Longleftrightarrow g$ parabolic or the identity,
- $f(g) < 2 \Longleftrightarrow g$ elliptic,
- $f(g_1) = f(g_2) \Longleftrightarrow k(g_1) = k(g_2) \Longleftrightarrow g_1$ and g_2 are conjugate transformations.

Let (g, h) be a pair of hyperbolic transformations fixing the upper half–plane U. Suppose that g and h have no common fixed points and denote

- $t = (r(g), r(h), a(h), a(g))$,
- $k_1 = k(g)$,
- $k_2 = k(h)$,
- $k_3 = k(g \circ h)$.

In order to derive an expression for $f(k_3)$ in terms of t, k_1 and k_2 we normalize by conjugation such that $r(h) = 1$, $a(h) = 0$ and $a(g) = \infty$. Then $t = r(g)$ and we have by formulae (1.4)

$$g(z) = k_1 z - t(k_1 - 1),$$

$$h(z) = \frac{z}{(1-k_2)z + k_2}.$$

It follows that

$$f(k_3) = f(g \circ h) = |tf(k_1k_2) + (1-t)f(k_1/k_2)|. \qquad (1.15)$$

Lemma 1.3.2 *The multipliers k_1, k_2, k_3 and $k_4 = k(g^2 \circ h)$ determine t and hence also the conjugacy class of (g,h) uniquely. If only k_1, k_2 and k_3 are fixed, then t has two possible alternatives.*

Proof. Retaining the above normalization we have by (1.15) either

$$r(g) = t = \frac{f(k_3) - f(k_1/k_2)}{f(k_1k_2) - f(k_1/k_2)}$$

or

$$r(g) = t' = \frac{f(k_3) + f(k_1/k_2)}{f(k_1/k_2) - f(k_1k_2)}.$$

Then $t - t' > 0$. By eliminating $f(k_3)$ we get

$$t + t' = -2\frac{k_1 + k_2}{(k_1-1)(k_2-1)}.$$

Similarly, replacing g by g^2 we get two values $\tilde{t}$ and $\tilde{t}'$ for $r(g) = r(g^2)$ satisfying

$$\tilde{t} + \tilde{t}' = -2\frac{k_1^2 + k_2}{(k_1^2-1)(k_2-1)}.$$

Since $k_2 > 1$, the function $k \mapsto (k+k_2)/(k-1)(k_2-1)$ is strictly decreasing for $k > 1$,and we have $t+t' < \tilde{t}+\tilde{t}'$. Therefore, the sets $\{t, t'\}$ and $\{\tilde{t}, \tilde{t}'\}$ can share at most one point. On the other hand, $r(g)$ belongs to the intersection of the sets $\{t, t'\}$ and $\{\tilde{t}, \tilde{t}'\}$. Hence $t = r(g)$ is uniquely determined. □

Lemma 1.3.3 *$k(g^{m_i} \circ h^{n_i}) \to \infty$ whenever* $\min(m_i, n_i) \to \infty$.

Proof. By (1.15)

$$f(g^{m_i} \circ h^{n_i}) = f(k_1^{m_i}k_2^{n_i}) \left| t + (1-t)\frac{k_1^{-m_i} + k_2^{-n_i}}{1 + k_1^{-m_i}k_2^{-n_i}} \right|.$$

Since g and h have no common fixed points, we have $t \neq 0$, and the assertion follows. □

It follows from Lemma 1.3.3 that $g^n \circ h^n$ is hyperbolic for sufficiently large values of n.

Lemma 1.3.4 *$a(g^n \circ h^n) \to a(g)$ and $r(g^n \circ h^n) \to r(h)$ as $n \to \infty$.*

Proof. We may suppose that the fixed points of g and h are finite. Choose disjoint closed intervals $I_1 \subset \mathbf{R}$ and $I_2 \subset \mathbf{R}$ containing $a(g)$ and $a(h)$ as interior points, respectively, but not containing $r(g)$ or $r(h)$. Choose n_0 such that $g^n(I_2) \subset I_1$ and $h^n(I_1) \subset I_2$ for $n \geq n_0$. Then $g^n(h^n(I_1)) \subset I_1$ and it follows that $a(g^n \circ h^n) \in I_1$ for $n \geq n_0$. Since $r(g^n \circ h^n) = a(h^{-n} \circ g^{-n})$, it follows similarly that $r(g^n \circ h^n) \to r(h)$ as $n \to \infty$. □

In the next lemma we suppose that g and h share at least one fixed point.

Lemma 1.3.5 *If $r(g) = a(h)$, then there are indices $m_i \to \infty$ and $n_i \to \infty$ such that $k(g^{m_i} \circ h^{n_i})$ stays bounded as $i \to \infty$.*

Proof. If $a(g) \neq r(h)$, then (1.15) remains valid with $t = 0$, and we have

$$f(g^{m_i} \circ h^{n_i}) = f(k_1^{m_i} / k_2^{n_i}). \tag{1.16}$$

If $a(g) = r(h)$ then $g^{m_i} \circ h^{n_i}$ is conjugate to $z \mapsto (k_1^{m_i} / k_2^{n_i}) z$ and (1.16) holds also in this case. Since $k_1 > 1$ and $k_2 > 1$, the assertion follows. □

In the following theorems, we consider groups G and G' of Möbius transformations acting in the upper half-plane U. An isomorphisms $j : G \to G'$ is *induced* by a Möbius transformation ψ if $j(g) = \psi \circ g \circ \psi^{-1}$ for all $g \in G$. Note that ψ is not uniquely determined by j. In fact ψ and $\overline{\psi}$ both induce j. Hence ψ can always be chosen such that $\psi(U) = U$. On the other hand, an orientation preserving ψ inducing j may map U onto the lower half-plane. The isomorphism j is *type-preserving* if g and $j(g)$ are of the same type for all $g \in G$.

Theorem 1.3.6 *Suppose that G is generated by a finite or countably infinite set $E = \{g_1, g_2, \ldots\}$ of hyperbolic transformations. Suppose that g_n and g_m share no fixed points for any $n \neq m$. Let $j : G \to G'$ be a type-preserving isomorphisms. If $k(g) = k(j(g))$ for every hyperbolic $g \in G$, then j is induced by a Möbius transformation.*

Proof. If E contains only one element, then there is nothing to prove.

Let x_1, x_2, x_3, x_4 be distinct fixed points of E. Suppose that $x_1 = r(h_1)$, $x_2 = r(h_2)$, $x_3 = a(h_3)$ and $x_4 = a(h_4)$ where either $h_i \in E$ or $h_i^{-1} \in E$, $i = 1, 2, 3, 4$.

To show that the points $y_1 = r(j(h_1))$, $y_2 = r(j(h_2))$, $y_3 = a(j(h_3))$ and $y_4 = a(j(h_4))$ are distinct, suppose, e.g., that $y_2 = y_3$. Then by Lemma 1.3.5, there exist indices $m_i \to \infty$ and $n_i \to \infty$ such that

$$k(j(h_2)^{m_i} \circ j(h_3)^{n_i}) \leq M < \infty \tag{1.17}$$

as $i \to \infty$. On the other hand, the points $x_2 = r(h_2)$ and $x_3 = a(h_3)$ are distinct. If we had $h_2 = h_3$, then the transformation $j(h_2) = j(h_3)$ would have only one fixed point and j could not be type–preserving. Hence h_2 and h_3 share no fixed points. Then by Lemma 1.3.3,

$$k(h_2^{m_i} \circ h_3^{n_i}) \to \infty$$

which contradicts (1.17).

We show next that

$$(x_1, x_2, x_3, x_4) = (y_1, y_2, y_3, y_4). \tag{1.18}$$

To that end, let $g_{1n} = h_3^n \circ h_1^n$ and $g_{2n} = h_4^n \circ h_2^n$. By Lemma 1.3.4,

$$r(g_{in}) \to x_i, \qquad r(j(g_{in})) \to y_i,$$

$$a(g_{in}) \to x_{i+2}, \qquad a(j(g_{in})) \to y_{i+2},$$

$i = 1, 2$. Consider the pair (g_{1n}, g_{2n}). By Lemma 1.3.3, $k(g_{1n}) \to \infty$ and $k(g_{2n}) \to \infty$. Since the cross–ratio of the fixed points of g_{1n} and g_{2n} is bounded away from 1, 0 and ∞, it follows by 1.15 that $g_{1n} \circ g_{2n}$ and $g_{1n}^2 \circ g_{2n}$ are hyperbolic for sufficiently large values of n. Then by Lemma 1.3.2, the pairs (g_{1n}, g_{2n}) and $(j(g_{1n}), j(g_{2n}))$ are conjugate, i.e.,

$$(r(g_{1n}), r(g_{2n}), a(g_{1n}), a(g_{2n})) = (r(j(g_{1n})), r(j(g_{2n})), a(j(g_{1n})), a(j(g_{2n}))),$$

from which (1.18) follows by letting $n \to \infty$.

Finally, keep the points x_2, x_3 and x_4 fixed and let ψ be the orientation preserving Möbius transformation defined by $\psi(x_i) = y_i$, $i = 2, 3, 4$. Let $x_1 = a(h_1')$ be any point distinct from x_2, x_3 and x_4 such that either $h_1' \in E$ or $(h_1')^{-1} \in E$. Then by (1.18), $\psi(x_1) = a(j(h_1'))$, and we have for all $g_i \in E$

$$\begin{aligned}
\psi(a(g_i)) &= a(j(g_i)), \\
\psi(r(g_i)) &= r(j(g_i)), \\
k(g_i) &= k(j(g_i)).
\end{aligned}$$

Hence $j(g_i) = \psi \circ g_i \circ \psi^{-1}$ for all generators $g_i \in E$, and the assertion follows. $\square$

For a Möbius group G acting in a disk or a half-plane, let G_+ denote the subgroup of the orientation preserving elements of G. For a moment, we restrict ourselves to countable groups G satisfying the following condition: For every $z \in \hat{\mathbf{C}}$, there exits $g_z \in G_+$ such that

$$\{ g \in G_+ \mid g(z) = z \} = \{ g_z^n \mid n = 0, \pm 1, \pm 2, \ldots \}. \tag{1.19}$$

For example, all Fuchsian groups satisfy these conditions. The last condition states that the stabilizer $\{ g \in G_+ \mid g(z) = z \}$ of $z \in \hat{\mathbf{C}}$ is a maximal cyclic subgroup of G_+ whenever $g_z \neq \mathrm{id}$.

Theorem 1.3.7 *Let G and G' be countable Möbius groups acting in U. Suppose that G satisfies (1.19) and has at least four distinct hyperbolic fixed points. Let $j : G \to G'$ be a type-preserving isomorphism. If $k(g) = k(j(g))$ for every hyperbolic $g \in G$, then j is induced by a Möbius transformation.*

Proof. Let the set $E = \{g_1, g_2, \ldots\}$ contain exactly one generator of every maximal hyperbolic cyclic subgroup of G. Let G_0 be the group generated by E. Then, by Theorem 1.3.6, $j \mid G_0$ is induced by a Möbius transformation ψ, i.e., $j(g) = \psi \circ g \circ \psi^{-1}$ for all $g \in G_0$.

If $g \in G \setminus G_0$, then g is either parabolic, elliptic, a reflection or a glide–reflection. For a glide–reflection $g \in G$, g^2 is hyperbolic. Hence

$$j(g)^2 = j(g^2) = \psi \circ g^2 \circ \psi^{-1} = (\psi \circ g \circ \psi^{-1})^2.$$

Since a glide–reflection is uniquely determined by its square, we have $j(g) = \psi \circ g \circ \psi^{-1}$.

Suppose that $g \in G \setminus G_0$ is not a glide–reflection. Choose a hyperbolic $h \in G$ such that $g(a(h)) \neq r(h)$ (such an h exists by Lemma 1.3.4). Then, by Lemma 1.3.1, $g \circ h^n$ is hyperbolic or a glide–reflection for sufficiently large values of n. Then

$$\begin{aligned} j(g) \circ j(h^n) &= j(g \circ h^n) = \psi \circ g \circ h^n \circ \psi^{-1} \\ &= \psi \circ g \circ \psi^{-1} \circ \psi \circ h^n \circ \psi^{-1} = \psi \circ g \circ \psi^{-1} \circ j(h^n) \end{aligned}$$

and hence $j(g) = \psi \circ g \circ \psi^{-1}$. $\square$

Corollary 1.3.8 *Let G and G' be countable Möbius groups acting in U. Suppose that G satisfies (1.19) and has at least four distinct hyperbolic fixed points. Let $j : G \to G'$ be a type-preserving isomorphism. If $j(g) = g$ for all $g \in G_+$, then $G = G'$ and $j = \mathrm{id}$.*

It follows from the Corollary 1.3.8 that it suffices in many cases to consider Möbius groups containing only orientation preserving elements. It is an interesting exercise to enumerate all groups having at most two hyperbolic fixed points.

1.4 Parametrization problem and classes $\mathcal{H}$, $\mathcal{P}$ and $\mathcal{E}$

The motivation of considering isomorphisms of Möbius groups has its roots in the theory of Teichmüller spaces. However, in the Teichmüller theory, only isomorphisms $j : G_1 \to G_2$ induced by quasiconformal mappings are considered. This gives rise to the following definition: For $k = 1, 2$, let G_k be a Möbius group acting in a disk or half-plane D_k. Denote by K_k the boundary of D_k. An isomorphism $j : G_1 \to G_2$ is *geometric* (on K_1) if

- j is type-preserving,
- there exists a homeomorphism $\psi : K_1 \to K_2$ inducing j on K_1.

It follows that ψ maps the hyperbolic or parabolic fixed points of G_1 onto the respective fixed points of G_2. Hence j preserves the cyclic order of the fixed points of G_1 on K_1. In the following, we consider only *principal-circle groups*, i.e., Möbius groups acting in a disk or half-plane.

Fix a principal-circle group G_0 and denote by $J(G_0)$ the set of all pairs (j, G) where G is a principal-group and $j : G_0 \to G$ is a geometric isomorphism.

Parametrization problem. Find a set $A \subset G_0$ having the following property: If $(j_1, G_1) \in J(G_0)$ and $(j_2, G_2) \in J(G_0)$, then $j_2 \circ j_1^{-1} : G_1 \to G_2$ is induced by a Möbius transformation if and only if

$$k(j_1(g)) = k(j_2(g))$$

for every $g \in A$. The set A *parametrizes* $J(G_0)$ and the numbers $k(j(g))$, $g \in A$, are *coordinates* of $(j, G) \in J(G_0)$.

In view of Theorem 1.3.7, we restrict ourselves to principal-circle groups generated by hyperbolic elements. If, firstly, the group G_0 is generated by one hyperbolic transformation g, then $A = \{g\}$ parametrizes $J(G_0)$. Secondly, let G_0 be generated by a pair (g, h) of hyperbolic transformations sharing no fixed points. By Lemma 1.3.2, the set

$$A = \{ g, h, g \circ h, g^2 \circ h \}$$

parametrizes $J(G_0)$. However, the fourth coordinate $k'_4 = k(j(g^2 \circ h))$ of (j, G) is in a special position: As soon as the first three coordinates

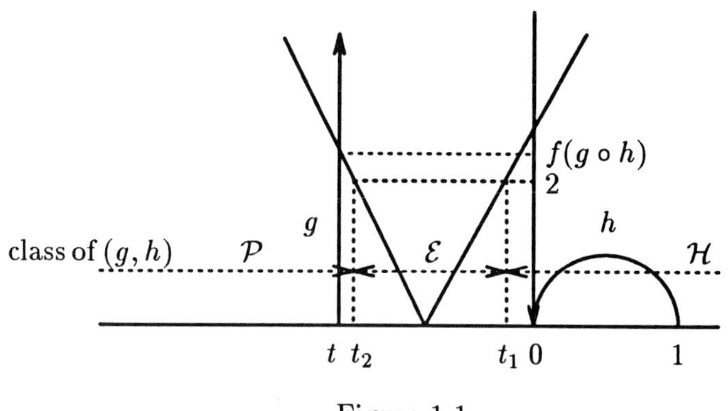

Figure 1.1:

$k_1' = k(j(g))$, $k_2' = k(j(h))$ and $k_3' = k(j(g \circ h))$ are fixed, k_4' has only two possible alternatives. We shall prove in Theorem 1.4.3 that k_4' is an almost superfluous coordinate for a geometric isomorphism j.

We call (g, h) *a principal–circle pair* if it satisfies the following conditions:

- g and h are hyperbolic transformations sharing no fixed points,
- the cross–ratio $t = (r(g), r(h), a(h), a(g))$ is real.

Groups generated by principal–circle pairs (g, h) are building blocks of all principal–circle groups. To our purposes it is necessary to limit ourselves to cases where $k_1 = k(g)$, $k_2 = k(h)$ and $k_3 = k(g \circ h)$ determine $k_4 = k(g^2 \circ h)$ uniquely. To that end, it suffices that $t = (r(g), r(h), a(h), a(g))$ will be uniquely determined. This in turn will be achieved by dividing the pairs (g, h) into three disjoint *classes* $\mathcal{H}$, $\mathcal{P}$ and $\mathcal{E}$ (cf. formula (1.15)):

$$\begin{aligned}
(g,h) \in \mathcal{H} &\iff f(g \circ h) = tf(k_1k_2) + (1-t)f(k_1/k_2) \geq 2 \\
&\iff t \geq t_1 = \frac{2 - f(k_1/k_2)}{f(k_1k_2) - f(k_1/k_2)}, \\
(g,h) \in \mathcal{P} &\iff f(g \circ h) = -tf(k_1k_2) - (1-t)f(k_1/k_2) \geq 2 \\
&\iff t \leq t_2 = \frac{-2 - f(k_1/k_2)}{f(k_1k_2) - f(k_1/k_2)}, \\
(g,h) \in \mathcal{E} &\iff t_2 < t < t_1.
\end{aligned}$$

Here $\mathcal{H}$ stands for "handle", $\mathcal{P}$ for "pants" and $\mathcal{E}$ for "elliptic". Since (g, h) is a principal–circle pair, $g \circ h$ is either hyperbolic, parabolic or elliptic. Note that $g \circ h$ is elliptic if and only if $(g, h) \in \mathcal{E}$.

Normalize the pair (g, h) by conjugation such that

$$\begin{aligned}
a(h) = 0 \quad &\text{and} \quad r(h) = 1, \\
a(g) = \infty \quad &\text{and} \quad r(g) = t.
\end{aligned}$$

In Figure 1.1, the axes of g and h are drawn in the same $z = t + iu$ plane with the graph of

$$u(t) = f(g \circ h) = |tf(k_1k_2) + (1-t)f(f_1/k_2)|. \tag{1.20}$$

Then $f(g \circ h)$ is the ordinate of the intersection point of (1.20) and the axis of g whereas the abscissa t of the intersection point gives the class of (g, h). In the class $\mathcal{H}$, $t = r(g)$ omits the points $0 = a(h)$ and $1 = r(h)$. Therefore, $f(g \circ h)$ omits the values $f(k_1k_2)$ and $f(k_1/k_2) = f(k_2/k_1)$ in $\mathcal{H}$. Since f maps the interval $[1, \infty]$ injectively onto $[2, \infty]$, we have, by Figure 1.1, the following lemma:

Lemma 1.4.1 *For any $k_1 > 1$, $k_2 > 1$ and $k_3 \geq 1$ there exists an up to conjugation unique pair $(g, h) \in \mathcal{P}$ such that*

$$k(g) = k_1, \quad k(h) = k_2, \quad k(g \circ h) = k_3. \tag{1.21}$$

For a pair $(g, h) \in \mathcal{H}$, $(k_3 - k_1k_2)(k_3 - k_1/k_2)(k_3 - k_2/k_1) \neq 0$. If $k_1 > 1$, $k_2 > 1$ and $k_3 \geq 1$ satisfy this condition, then there exists an up to conjugation unique pair $(g, h) \in \mathcal{H}$ satisfying (1.21). □

In subsequent applications of the above lemma, we shall exclude the class $\mathcal{E}$ and consider only principal–circle pairs (g, h) with non–elliptic $g \circ h$. If we then know that the class of (g, h) is fixed, then k_1, k_2 and k_3 determine the conjugacy class of (g, h). Hence we need not know whether (g, h) is actually in $\mathcal{P}$ or in $\mathcal{H}$, it suffices to know the invariance of the class only.

We show next that the invariance of the classes $\mathcal{P}$ and $\mathcal{H}$ is pertinent to the parametrization problem. We consider first the classes $\mathcal{H}$, $\mathcal{P}$ and $\mathcal{E}$ in detail.

Considering the cross–ratios

$$t = (r(g), r(h), a(h), a(g)) = (t, 1, 0, \infty)$$

and

$$1 - t = (r(g^{-1}), r(h), a(h), a(g^{-1})) = (\infty, 1, 0, t)$$

associated with the pairs (g, h) and (g^{-1}, h), respectively, we see that

- the pairs (g, h), (h, g), (g^{-1}, h^{-1}) and (h^{-1}, g^{-1}) are in the same class,
- the pairs (g^{-1}, h), (h, g^{-1}), (g, h^{-1}) and (h^{-1}, g) are in the same class.

For a moment, we normalize by conjugation such that g and h map the unit disk D onto itself. Then g and h have well–defined *isometric circles* $I(g)$ and $I(h)$. In general, $I(g)$ is defined by the following properties (Figure 1.2):

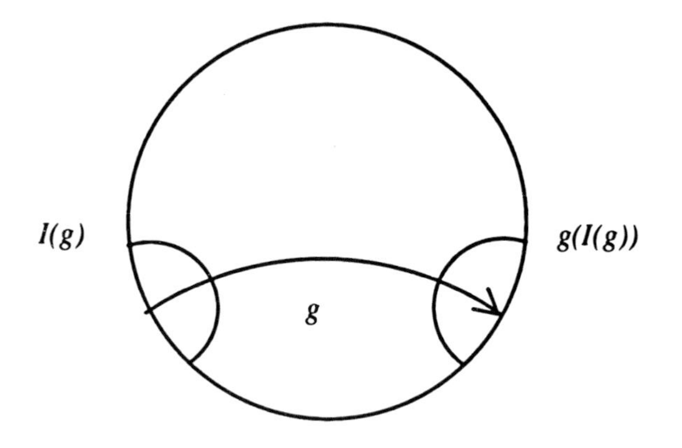

Figure 1.2:

- $I(g)$ is perpendicular to $ax(g)$ and to the unit circle K,
- $g(I(g))$ and $I(g)$ have the same radius.

It follows that $g(I(g)) = I(g^{-1})$ and g maps the inside of $I(g)$ onto the outside of $I(g^{-1})$.

Since g and h share no fixed points, we may suppose that

$$\begin{aligned} a(h) &= 1 \quad \text{and} \quad r(h) = -1, \\ a(g) &= e^{i\vartheta} \quad \text{and} \quad r(g) = -e^{\pm i\vartheta}, \qquad 0 < \vartheta < \pi. \end{aligned}$$

In Figure 1.3, three different alternatives 3.1–3.3 for the cyclic order of the fixed points of g and h are represented. Here 3.1 contains the whole classes $\mathcal{P}$ and $\mathcal{E}$ and possibly a part of $\mathcal{H}$. The rest of the class $\mathcal{H}$ is contained in 3.2 and 3.3, cf. Figure 1.1.

The case 3.1 has five different subcases 3.1.1–3.1.5 which are found by drawing the isometric circles of g, h, g^{-1} and h^{-1}.

3.1.1.

- $t < t_2$,
- $(g, h) \in \operatorname{Int} \mathcal{P}$, the *interior* of $\mathcal{P}$,
- $I(h)$ and $I(g^{-1})$ are exterior to each other.

3.1.2.

- $t = t_2$,
- $(g, h) \in \operatorname{Bd} \mathcal{P}$, the *boundary* of $\mathcal{P}$,

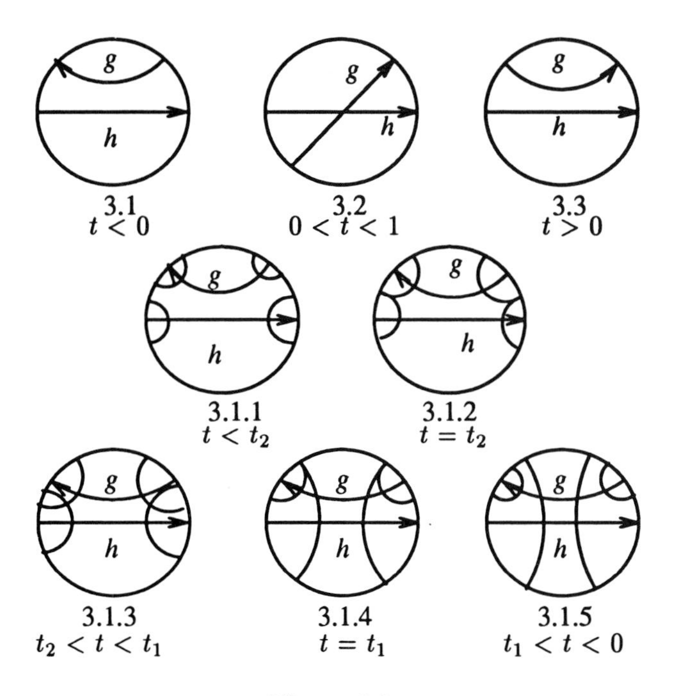

Figure 1.3:

- $I(h)$ is exterially tangent to $I(g^{-1})$,
- the tangential point x is fixed by the parabolic transformation $g \circ h$.

3.1.3.

- $t_2 < t < t_1$,
- $(g, h) \in \mathcal{E}$,
- $I(h)$ and $I(g^{-1})$ intersect,
- the intersection point x is fixed by the elliptic transformation $g \circ h$.

3.1.4.

- $t = t_1$,
- $(g, h) \in \operatorname{Bd} \mathcal{H}$, the *boundary* of $\mathcal{H}$,
- $I(h)$ and $I(g^{-1})$ are interially tangent to each other,
- the tangential point is fixed by the parabolic transformation $g \circ h$.

3.1.5.

- $t_1 < t < 0$,
- $(g, h) \in \operatorname{Int} \mathcal{H}$, the *interior* of $\mathcal{H}$,
- one of the circles $I(h)$ and $I(g^{-1})$ is interior to another.

The interior of $\mathcal{H}$ consists of the topologically different subclasses 3.1.5, 3.2 and 3.3. Pairs with intersecting axes represent *real handles.* Therefore, the subclass 3.2 of $\mathcal{H}$ is denoted by $\mathcal{H}_R$.

After this detailed listing of different alternatives we can prove that the classes $\mathcal{H}$, $\mathcal{P}$ and $\mathcal{E}$ are invariant under geometric isomorphisms. The treatment of the parametrization problem will be based on this result.

Theorem 1.4.2 *Let G_0 and G be principal-circle groups, $j : G_0 \to G$ a geometric isomorphism and g, $h \in G_0$ hyperbolic elements sharing no fixed points. Then the pairs (g, h) and $(j(g), j(h))$ are both in the same class $\mathcal{H}$, $\mathcal{P}$ or $\mathcal{E}$.*

Proof. Denote $g' = j(g)$ and $h' = j(h)$. Let t', t'_1 and t'_2 be the parameters associated with the pair (g', h').

Suppose that $t < t_2$. Since j preserves the cyclic order of the fixed points and the types of the transformations, we have, by 3.1.1.–3.1.5, either $t' < t'_2$ or $t'_1 < t' < 0$. In case 3.1.1, the fixed points $a(g)$ and $r(g)$ bound an arc of the unit circle not containing points x' and y' such that $h(x') = y'$. Similarly, $a(h)$ and $r(h)$ bound an arc not containing points x' and y' such that $g(x') = y'$. Hence the cases 3.1.1 and 3.1.5 cannot be geometrically isomorphic, and it follows that $t' < t'_2$.

Suppose that $t = t_2$. Since j is type–preserving and $g \circ h$ is parabolic, we have either $t' = t'_2$ or $t' = t'_1$. In cases 3.1.2 and 3.1.4 the cyclic orders of the fixed points of g, h and $g \circ h$ are different. Hence $t' = t'_2$. Since we can apply the same reasoning to j^{-1}, we have proved that $(g, h) \in \mathcal{P}$ if and only if $(g', h') \in \mathcal{P}$. Moreover, since j is type–preserving, $(g, h) \in \mathcal{E}$ if and only if $(g', h') \in \mathcal{E}$, and the assertion follows. □

We can now show that the fourth coordinate k'_4 is superfluous if G_0 is generated by a pair of the class $\mathcal{P}$ or $\mathcal{H}$.

Theorem 1.4.3 *Let the Möbius group G_0 be generated by a principal-circle pair (g, h). If $g \circ h$ is non–elliptic, then $A = \{ g, h, g \circ h \}$ parametrizes $J(G_0)$.*

Proof. Let $(j_1, G_1) \in J(G_0)$ and $(j_2, G_2) \in J(G_0)$. Since $g \circ h$ is non–elliptic, (g, h) is in $\mathcal{P} \cup \mathcal{H}$. By Theorem 1.4.2, the pairs $(j_1(g), j_1(h))$ and

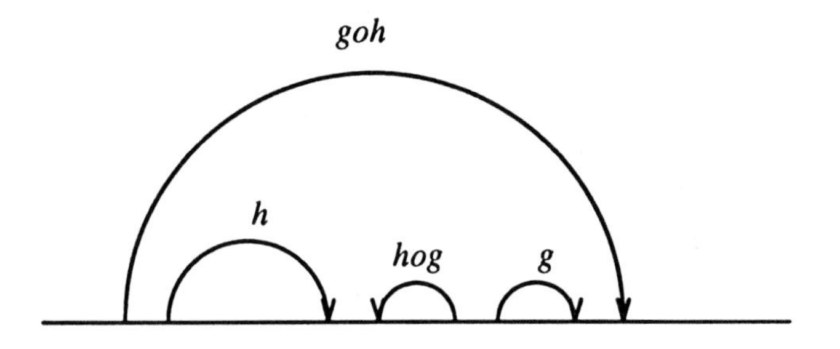

Figure 1.4:

$(j_2(g), j_2(h))$ are in the same class $\mathcal{P}$ or $\mathcal{H}$. Suppose that $k(j_1(g) = k(j_2(g))$, $k(j_1(h)) = k(j_2(h))$ and $k(j_1(g \circ h)) = k(j_2(g \circ h))$. Then, by Lemma 1.4.1, the pairs are conjugate. Since they generate G_1 and G_2, respectively, $j_2 \circ j_1^{-1} : G_1 \to G_2$ is induced by a Möbius transformation. □

1.5 Geometrical properties of the classes $\mathcal{P}$ and $\mathcal{H}$

For later reference, we consider geometrical properties of the principal–circle pairs (g, h). Especially, we study the *commutator* $[g, h] = h \circ g^{-1} \circ h^{-1} \circ g$ of g and h.

Lemma 1.5.1 *If $(g, h) \in \operatorname{Int} \mathcal{P}$, then $g \circ h$ and $h \circ g$ are hyperbolic, the axes of g, h, $g \circ h$ and $h \circ g$ are disjoint in pairs, and the cyclic order of the fixed points is given by Figure 1.4.*

Proof. Normalize by 3.1.1. Since the pairs (g, h) and (h, g) are both in $\operatorname{Int} \mathcal{P}$, $g \circ h$ and $h \circ g$ are hyperbolic by Figure 1.1. Let L be the non–Euclidean line in D perpendicular to $I(h)$ and $I(g^{-1})$, see Figure 1.5.

By symmetry, $h(L) = g^{-1}(L)$ is perpendicular to $I(g)$ and $I(h^{-1})$. Since

$$(g \circ h)(L) = g(g^{-1}(L)) = L$$

and

$$(h \circ g)(g^{-1}(L)) = h(L) = g^{-1}(L),$$

we have $L = ax(g \circ h)$ and $g^{-1}(L) = ax(h \circ g)$, and the assertions follows. □

To study the commutator of g and h, suppose that the axes of g and h intersect, i.e., $(g, h) \in \mathcal{H}_R$.

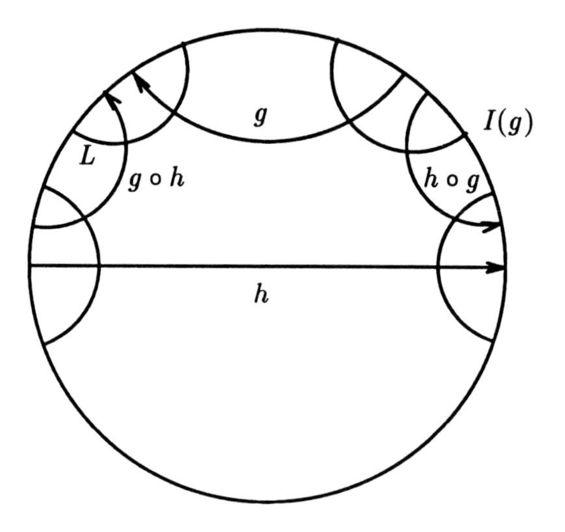

Figure 1.5:

Lemma 1.5.2 *For the pair $(g,h) \in \mathcal{H}_R$, the commutator $c = [g,h]$ is hyperbolic if and only if $(h, g^{-1} \circ h^{-1} \circ g) \in \operatorname{Int} \mathcal{P}$.*

Proof. Denote $h' = g^{-1} \circ h^{-1} \circ g$. To consider the class of the pair (h,h'), let $t = (r(h), r(h'), a(h'), a(h))$ and $k = k(h) = k(h')$. Then

$$\begin{aligned} t_1 &= \frac{2 - f(k/k)}{f(k^2) - f(k/k)} = 0, \\ t_2 &= \frac{-2 - f(k/k)}{f(k^2) - f(k/k)} = \frac{-4}{f(k^2) - 2} < 0. \end{aligned}$$

We may suppose that $a(h) = \infty$, $a(h') = 0$, $r(h') = 1$ and $r(h) = t$. Since

$$\begin{aligned} g^{-1}(r(h)) &= a(h') = 0, \\ g^{-1}(a(h)) &= r(h') = 1, \end{aligned}$$

we have $g(0) = t$ and $g(1) = \infty$. On the other hand, the axis of g intersects with the axes of h and h'. From $g(1) = \infty$ it then follows that $a(g) < t < 0 < r(g) < 1$, as depicted in Figure 1.6.

The commutator $c = h \circ h'$ is hyperbolic if and only if $(h,h') \in \operatorname{Int} \mathcal{P} \cup \operatorname{Int} \mathcal{H}$, i.e., if and only if $t < t_2$ or $t_1 < t$. Since $t_1 = 0$ and $t < 0$, the assertion follows. □

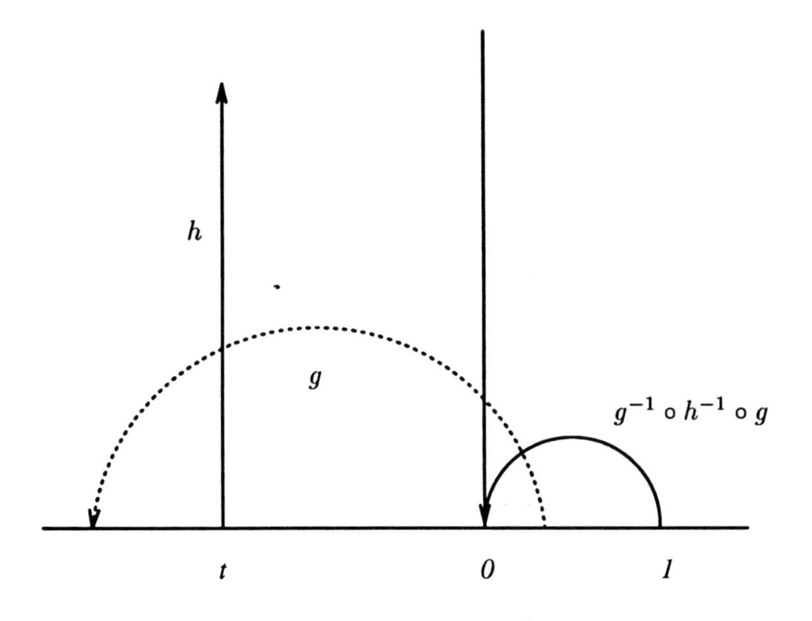

Figure 1.6:

Suppose that $(g, h) \in \mathcal{H}_R$ and that $c = [g, h]$ is hyperbolic. Then $(h, g^{-1} \circ h^{-1} \circ g) \in \operatorname{Int} \mathcal{P}$, and we obtain, by Lemma 1.5.1, the following Figure 1.7 showing how the axes of g, h, $c = h \circ g^{-1} \circ h^{-1} \circ g$ and $g^{-1} \circ h^{-1} \circ g \circ h$ are located.

We consider next what can be said about the pair $(g, h) \in \mathcal{H}_R$ if the multipliers $k(g)$, $k(h)$ and $k(c)$ are known.

Theorem 1.5.3 *Let $(g, h) \in \mathcal{H}_R$ be a pair with hyperbolic $c = [g, h]$. If the multipliers $k_1 = k(g)$, $k_2 = k(h)$ and $k_4 = k(c)$ are known, then $k_3 = k(g \circ h)$ has two possible values except in the case*

$$f(k_4) + 2 = \frac{1}{4}(f(k_1^2) - 2)(f(k_2^2) - 2) \tag{1.22}$$

when k_3 is uniquely determined.

Proof. A lengthy but straightforward calculation yields

$$f(k_4) = |t(1 - t)(f(k_1^2) - 2)(f(k_2^2) - 2) - 2|$$

with $t = (r(g), r(h), a(h), a(g)), 0 < t < 1$. Since $f(k_4) > 2$ and $t(1 - t) > 0$, we have in fact

$$f(k_4) = t(1 - t)(f(k_1^2) - 2)(f(k_2^2) - 2) - 2. \tag{1.23}$$

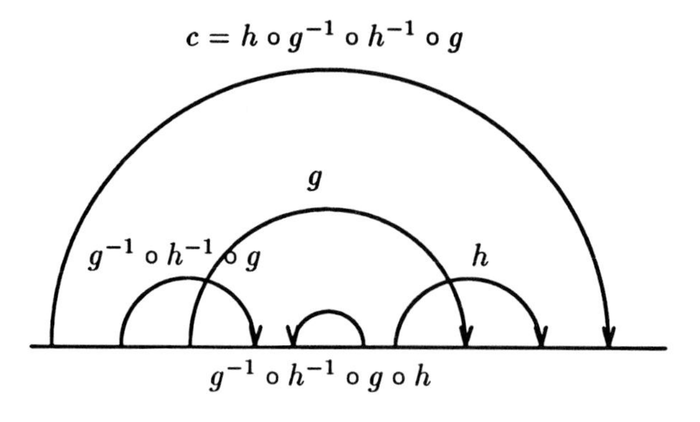

Figure 1.7:

On the other hand,

$$f(k_3) = tf(k_1k_2) + (1-t)f(k_1/k_2).$$

Eliminating t we then obtain

$$t(1-t) = \frac{f(k_4)+2}{(f(k_1^2)-2)(f(k_2^2)-2)} = \frac{(f(k_3)-f(k_1/k_2))(f(k_1k_2)-f(k_3))}{(f(k_1k_2)-f(k_1/k_2))^2}. \tag{1.24}$$

With respect to $f(k_3)$, the equation is of the second degree. Hence k_3 has at most two possible values. The last assertion follows by considering the discriminant of the equation. □

Corollary 1.5.4 *For the pair $(g,h) \in \mathcal{H}_R$ with hyperbolic $c = [g,h]$, the axes of g and h are perpendicular if and only if (1.22) holds. In other cases k_1, k_2 and k_4 determine the acute angle between the axes of g and h uniquely.*

Proof. By (1.23), the case (1.22) occurs if and only if $t = 1/2$. This in turn occurs, by Figure 1.1, if and only if the axes of g and h are perpendicular.

Suppose that $t \neq 1/2$ and denote by k_3 and k_3' the two values of $k(g \circ h)$. If

$$\begin{aligned} f(k_3) &= tf(k_1k_2) + (1-t)f(k_1/k_2), \quad 0 < t < 1, \\ f(k_3') &= t'f(k_1k_2) + (1-t')f(k_1/k_2), \quad 0 < t' < 1, \end{aligned}$$

then by (1.24), $t(1-t) = t'(1-t')$. Hence $t' = 1-t$. The assertion follows by Figure 1.1. □

Corollary 1.5.5 *Let $(g,h) \in \mathcal{H}_R$ be a pair with hyperbolic $c = [g,h]$. Then*

$$\frac{4k_2}{(k_2-1)^2} < \frac{(k_1-1)^2}{4k_1}. \tag{1.25}$$

Especially, if $k_2 < 1+\varepsilon$, then $k_1 > 16/\varepsilon^2$.

Proof. In (1.23) we have $0 < t(1-t) < 1/4$. Hence

$$2 < f(k_4) < \frac{1}{4}(f(k_1^2)-2)(f(k_2^2)-2)-2$$

and

$$16 < (k_1 - 2 + \frac{1}{k_1})(k_2 - 2 + \frac{1}{k_2}) = \frac{(k_1-1)^2}{k_1}\frac{(k_2-1)^2}{k_2}$$

from which (1.25) follows.

Finally

$$\frac{k_1}{4} > \frac{(k_1-1)^2}{4k_1} > \frac{4k_2}{(k_2-1)^2} > \frac{4(1+\varepsilon)}{\varepsilon^2} > \frac{4}{\varepsilon^2}. \quad \square$$

We call a hyperbolic $g \in G$ a *primary element* of the principal–circle group G if, for any $z \in ax(g)$, the interval of $ax(g)$ bounded by z and $g(z)$ contains no pair of points equivalent under G.

Let (g,h) be a principal–circle pair. Suppose that g and h have non–intersecting axes. Then, replacing g by g^{-1} if necessary, we may suppose that

$$t = (r(g), r(h), a(h), a(g)) < 0. \tag{1.26}$$

Theorem 1.5.6 *The pair (g,h) satisfying (1.26) is in* Int $\mathcal{P}$ *if and only if*

- *$g \circ h$ is hyperbolic,*
- *g and h are primary elements of the group $\langle g,h\rangle$ generated by g and h.*

Proof. By (1.26), the pair (g,h) satisfies exactly one of the cases 3.1.1–3.1.5. The transformation $g \circ h$ is hyperbolic if and only if either 3.1.1 or 3.1.5 occurs. If this is the case, then g and h are both primary elements of $\langle g,h\rangle$ if and only if 3.1.1 holds. $\square$

Normalize such that g and h fix the unit disk D. The *hyperbolic metric* d_D of D is defined by the line element

$$ds = \frac{2|dz|}{1-|z|^2}.$$

We have, for a hyperbolic transformation $g : D \to D$,

$$d_D(g(z), z) \geq \log k(g)$$

where the equality holds if and only if $z \in ax(g)$.

Hyperbolic distances are invariant under Möbius transformations fixing D, and geodesics of this metric are non–Euclidean lines. Let $d_D(L_1, L_2)$ denote the shortest distance between two non–Euclidean lines L_1 and L_2.

Theorem 1.5.7 *For the pair* $(g, h) \in \operatorname{Int}\mathcal{P}$, *let* $\log k = d_D(ax(g), ax(h))$, $k_1 = k(g)$ *and* $k_2 = k(h)$. *Then*

$$k > \frac{f(\sqrt{k_1 k_2}) + f(\sqrt{k_1/k_2})}{f(\sqrt{k_1 k_2}) - f(\sqrt{k_1/k_2})}.$$

Proof. Normalize by 3.1.1 and denote by iy the intersection point of $ax(g)$ with the imaginary axis. If $a(g) = e^{i\vartheta}$ and $r(g) = -e^{-i\vartheta}$, then

$$t = (-e^{-i\vartheta}, -1, 1, e^{i\vartheta}) = \frac{1 - \cos\vartheta}{2\cos\vartheta}.$$

Hence

$$-\cos\vartheta = \frac{1}{1 - 2t}.$$

Denote by L the circle determined by $ax(g)$. Since L and K are orthogonal, the point i/y lies on L. Hence L has the radius $r = \frac{1}{2}\left(\frac{1}{y} - y\right)$. By Figure 1.8,

$$\tan\alpha = -\cot\vartheta = r = \frac{1 - y^2}{2},$$

and it follows that

$$y = \sqrt{\frac{1 + \cos\vartheta}{1 - \cos\vartheta}} = \sqrt{\frac{-t}{1 - t}}.$$

Since $(g, h) \in \operatorname{Int}\mathcal{P}$, we have $t < t_2$ and

$$y > \sqrt{\frac{-t_2}{1 - t_2}} = \frac{f(\sqrt{k_1/k_2})}{f(\sqrt{k_1 k_2})}.$$

The assertion follows now from $k = (1 + y)/(1 - y)$. □

Corollary 1.5.8 *If, under the hypotheses of Theorem 1.5.7,* $k_j < 1 + \varepsilon < 2$, $j = 1, 2$, *then* $k > 8/\varepsilon^2$.

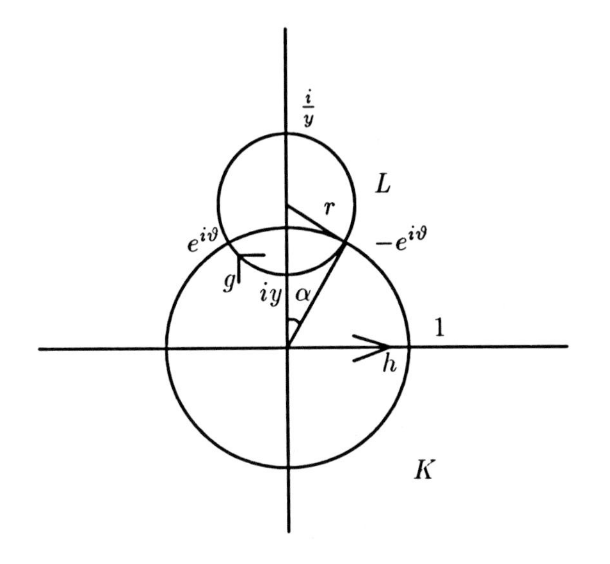

Figure 1.8:

Proof. For all ε, $0 < \varepsilon < 1$, we have

$$(1+\varepsilon)^{\frac{1}{2}} < 1 + \frac{1}{2}\varepsilon$$

and

$$(1+\varepsilon)^{-\frac{1}{2}} < 1 - \frac{1}{2}\varepsilon + \frac{3}{8}\varepsilon^2.$$

Hence

$$f(1+\varepsilon) < 2 + \frac{3}{8}\varepsilon^2.$$

Consequently

$$\begin{aligned} k &> \frac{f(\sqrt{k_1k_2}) + f(\sqrt{k_1/k_2})}{f(\sqrt{k_1k_2}) - f(\sqrt{k_1/k_2})} > \frac{f(\sqrt{k_1k_2}) + 2}{f(\sqrt{k_1k_2}) - 2} \\ &\geq \frac{f(1+\varepsilon)+2}{f(1+\varepsilon)-2} > \frac{4+\frac{3}{8}\varepsilon^2}{\frac{3}{8}\varepsilon^2} > \frac{8}{\varepsilon^2}. \square \end{aligned}$$

Let $(g_1, g_2) \in \operatorname{Int} \mathcal{P}$ be a pair of hyperbolic transformations fixing D. Denote

$$g_3 = g_2^{-1} \circ g_1^{-1}.$$

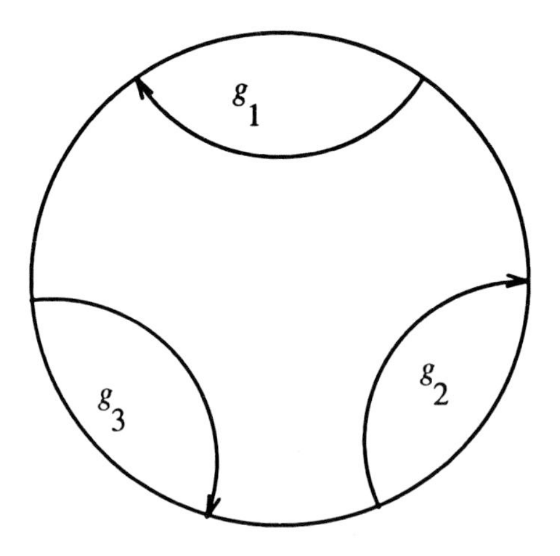

Figure 1.9:

Replacing g_1 and g_2 by g_1^{-1} and g_2^{-1} if necessary we may suppose, by Lemma 1.5.1, that the ordered triple of points $a(g_3)$, $a(g_2)$, $a(g_1)$ defines the positive orientation of the unit circle K, cf. Figure 1.9.

In the rest of this section, the normalization of Figure 1.9 is assumed to be fulfilled. Hence, when conjugating pairs $(g_1, g_2) \in \operatorname{Int} \mathcal{P}$, we may conjugate only by orientation preserving Möbius transformations fixing D.

The next lemma shows that g_1, g_2 and g_3 are in quite a symmetric position:

Lemma 1.5.9 *If* $(g_1, g_2) \in \operatorname{Int} \mathcal{P}$, *then* $(g_2, g_3) \in \operatorname{Int} \mathcal{P}$ *and* $(g_3, g_1) \in \operatorname{Int} \mathcal{P}$.

Proof. We show e.g. that $(g_2, g_3) \in \operatorname{Int} \mathcal{P}$. Since $g_3^{-1} \circ g_2^{-1} = g_1$ is hyperbolic, the pair (g_2, g_3) can be conjugated, by 3.1.1–3.1.5, such that either 10.1, 10.2 or 10.3 in Figure 1.10 holds.

In all cases the non–Euclidean line L perpendicular to $I(g_3)$ and to $I(g_2^{-1})$ is the axis of g_1 (cf. Figure 1.5). Comparing with Figure 1.9 we see that only the case 10.1 remains, i.e., $(g_2, g_3) \in \operatorname{Int} \mathcal{P}$. □

We show next that the class $\operatorname{Int} \mathcal{P}$ is, in a sense, transitive. Let g and ψ be hyperbolic transformations fixing D. Then generally

$$I(\psi \circ g \circ \psi^{-1}) \neq \psi(I(g)),$$

i.e., isometric circles do not behave nicely under conjugation. However, we have:

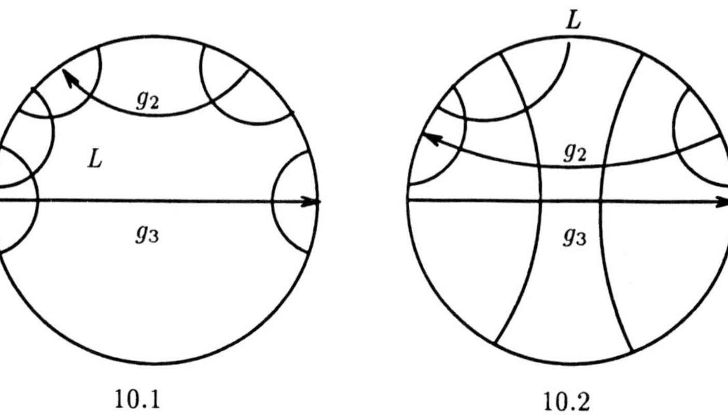

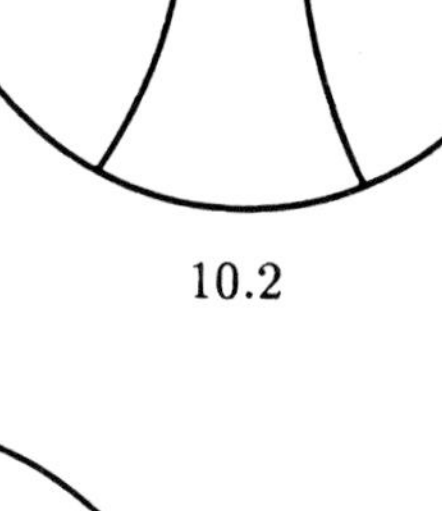

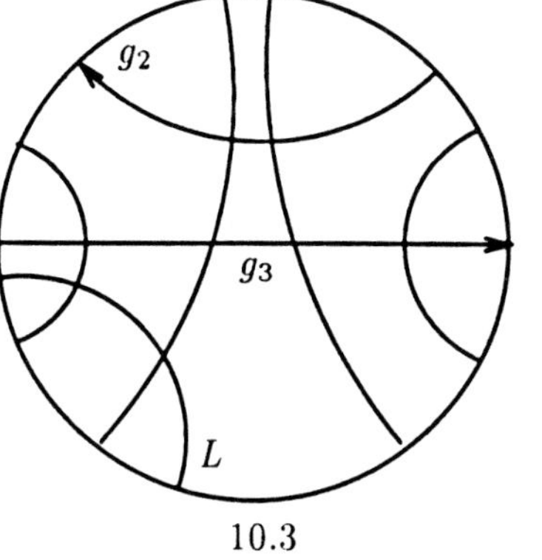

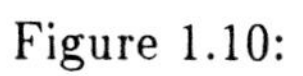

Figure 1.10:

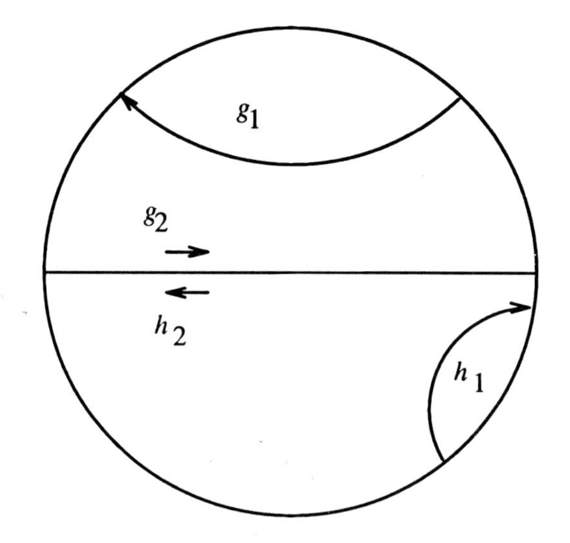

Figure 1.11:

Lemma 1.5.10 *If $ax(\psi)$ is the diameter of D perpendicular to $ax(g)$, then*

$$I(\psi \circ g \circ \psi^{-1}) = \psi(I(g)).$$

Proof. By symmetry, $\psi(I(g))$ and $\psi(I(g^{-1}))$ have the same Euclidean radius. Since they are in addition perpendicular to $ax(\psi \circ g \circ \psi^{-1}) = \psi(ax(g))$ and to K, the assertion follows. □

Theorem 1.5.11 *If $(g_1, g_2) \in \operatorname{Int} \mathcal{P}$, $(h_1, h_2) \in \operatorname{Int} \mathcal{P}$ and $g_2 = h_2^{-1}$, then $(g_1, h_1) \in \operatorname{Int} \mathcal{P}$.*

Proof. Conjugate such that $r(g_2) = -1$, $a(g_2) = 1$, $a(g_1) = e^{i\vartheta}$ and $r(g_1) = -e^{-i\vartheta}$, see Figure 1.11. By the normalization of Figure 1.9, $ax(h_1)$ lies then in the lower half–plane.

The non–Euclidean distances of the axes of g_1, g_2 and h_1 satisfy

$$d_D(ax(g_1), ax(h_1)) \geq d_D(ax(g_1), ax(g_2)) + d_D(ax(g_2), ax(h_1)), \qquad (1.27)$$

where the equality holds if and only if $ax(h_1)$ is perpendicular to the imaginary axis. These observations follow by considering the non–Euclidean line perpendicular to the axes of g_1 and h_1.

Let φ_j, $j = 1, 2$, be the hyperbolic transformation for which

- $\varphi_j(D) = D$,

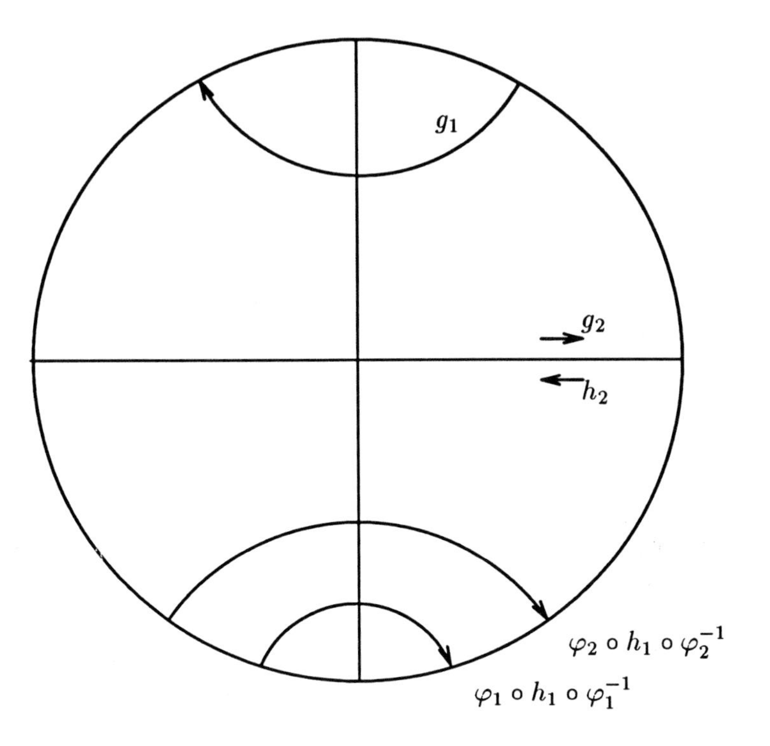

Figure 1.12:

- $ax(\varphi_j) = ax(g_j)$,
- the axis of $\varphi_j \circ h_1 \circ \varphi_j^{-1}$ is perpendicular to the imaginary axis.

By (1.27), Figure 1.12 holds. Since $(h_1, h_2) \in \operatorname{Int} \mathcal{P}$, the isometric circles of h_2^{-1} and $\varphi_2 \circ h_1 \circ \varphi_2^{-1}$ are exterior to each other by 3.1.1. By Figure 1.12, also $I(h_2^{-1})$ and $I(\varphi_1 \circ h_1 \circ \varphi_1^{-1})$ are exterior to each other. On the other hand, also $I(g_2) = I(h_2^{-1})$ and $I(g_1^{-1})$ are exterior to each other.

Let ψ be the hyperbolic transformation for which

- $\psi(D) = D$,
- $ax(\psi)$ is the imaginary axis,
- $r(\psi \circ \varphi_1 \circ h_1 \circ \varphi_1^{-1} \circ \psi^{-1}) = -1$,
- $a(\psi \circ \varphi_1 \circ h_1 \circ \varphi_1^{-1} \circ \psi^{-1}) = 1$,

Then, by Lemma 1.5.10, the isometric circles of $\psi \circ g_1 \circ \psi^{-1} = \psi \circ \varphi_1 \circ g_1 \circ \varphi_1^{-1} \circ \psi^{-1}$ and $\psi \circ \varphi_1 \circ h_1 \circ \varphi_1^{-1} \circ \psi^{-1}$ are exterior to each other and the assertion follows by 3.1.1. □

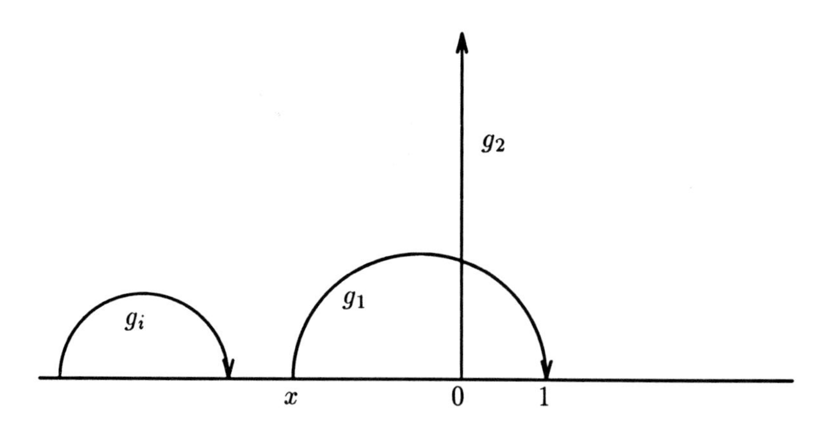

Figure 1.13:

1.6 Parametrization of principal–circle groups by multipliers

In this section, we consider principal–circle groups G acting in the upper half–plane U. Suppose that G is generated by a set

$$E = \{g_1, g_2, \ldots\}$$

of hyperbolic transformations allowing the following normalization (Figure 1.13):

$$\left.\begin{array}{l} a(g_1) = 1, \quad r(g_1) = x < 0, \\ a(g_2) = \infty, \quad r(g_2) = 0, \\ r(g_i) < a(g_i) < x, \quad i = 3, 4, \ldots . \end{array}\right\} \tag{1.28}$$

Lemma 1.6.1 *Let $E = \{g_1, g_2, \ldots\}$ be a set of hyperbolic transformations satisfying (1.28). If the classes of the pairs (g_i, g_1) and (g_i, g_2), $i = 3, 4, \ldots$, are fixed in $\mathcal{P} \cup \mathcal{H}$, then the numbers $k(g_i)$, $k(g_{i+1} \circ g_1)$ and $k(g_{i+2} \circ g_2)$, $i = 1, 2, \ldots$, determine E uniquely.*

Proof. Since $(g_2, g_1) \in \mathcal{H}$, the point $x = r(g_1)$ is uniquely determined by Lemma 1.4.1.

Choose $i \geq 3$ and let γ and γ' be two candidates for g_i satisfying

- $k(\gamma) = k(\gamma')$,
- $k(\gamma \circ g_1) = k(\gamma' \circ g_1)$,
- $k(\gamma \circ g_2) = k(\gamma' \circ g_2)$,

- (γ, g_1) and (γ', g_1) are in the same class $\mathcal{P}$ or $\mathcal{H}$,
- (γ, g_2) and (γ', g_2) are in the same class $\mathcal{P}$ or $\mathcal{H}$.

Then it suffices to show that $\gamma = \gamma'$.

Suppose that $\gamma \neq \gamma'$. By Lemma 1.4.1, the pairs (γ, g_1) and (γ', g_1) as well as the pairs (γ, g_2) and (γ', g_2) are conjugate. Since

$$r(\gamma) < a(\gamma) < x \quad \text{and} \quad r(\gamma') < a(\gamma') < x,$$

there are hyperbolic transformations ψ and σ such that

$$\gamma' = \sigma \circ \gamma \circ \sigma^{-1}, \quad \sigma(0) = 0, \quad \sigma(\infty) = \infty,$$
$$\gamma' = \psi \circ \gamma \circ \psi^{-1}, \quad \psi(1) = 1, \quad \psi(x) = x.$$

Let y_1 and y_2 denote the fixed points of γ. Then $\psi(y_1) = \sigma(y_1)$ and $\psi(y_2) = \sigma(y_2)$. Since $\sigma(z) = kz$ for $k = k(\sigma)$ or $k^{-1} = k(\sigma)$, we have

$$\frac{\psi(y_1)}{\psi(y_2)} = \frac{y_1}{y_2}.$$

Let generally, by formulae (1.4)–(1.6),

$$\psi(z) = \frac{(ak - r)z - ar(k-1)}{(k-1)z - kr + a}, \tag{1.29}$$

$k > 0, k \neq 1$, be the hyperbolic transformation with the real fixed points a and r and with the multiplier $\max(k, 1/k)$. If there exist real numbers y_1 and y_2 such that $y_2\psi(y_1) = y_1\psi(y_2)$, then, by direct calculation,

$$k = \frac{r(y_1 - a)(y_2 - a)}{a(y_1 - r)(y_2 - r)}. \tag{1.30}$$

If we insert $a = 1$, $r = x < 0$, $y_1 < x$ and $y_2 < x$, we obtain $k < 0$ which is impossible. Hence $\psi = \mathrm{id}$ and $\gamma = \gamma'$. □

Theorem 1.6.2 *Suppose that the principal-circle group G_0 is generated by a set $E_0 = \{g_1, g_2, \ldots\}$ of hyperbolic transformations satisfying (1.28). Suppose that the transformations $g_i \circ g_1$ and $g_i \circ g_2$ are non-elliptic for $i = 3, 4, \ldots$. Then $A = \{g_i, g_{i+1} \circ g_1, g_{i+2} \circ g_2 \mid i = 1, 2, \ldots\}$ parametrizes $J(G_0)$.*

Proof. Choose $(j_k, G_k) \in J(G_0), k = 1, 2$. Since $j_1 : G_0 \to G_1$ and $j_2 : G_0 \to G_2$ preserve the cyclic order of the real fixed points of G_0, the sets

$j_1(E_0)$ and $j_2(E_0)$ are conjugate to sets E_1 and E_2, respectively, for which conditions of the type (1.28) are valid. Suppose that

$$\begin{aligned} k(j_1(g_i)) &= k(j_2(g_i)), \\ k(j_1(g_{i+1} \circ g_1)) &= k(j_2(g_{i+1} \circ g_1)), \\ k(j_1(g_{i+2} \circ g_2)) &= k(j_2(g_{i+2} \circ g_2)) \end{aligned}$$

for $i = 1, 2, \ldots$. Since the pairs (g_i, g_1) and (g_i, g_2) are in $\mathcal{P} \cup \mathcal{H}$ for $i = 3, 4, \ldots$, we have, by Theorem 1.4.2 and Lemma 1.6.1, $E_1 = E_2$, and the assertion follows. □

Retaining the assumptions of Theorem 1.6.2, suppose that

$$E_0 = \{g_1, \ldots, g_s\}$$

is a finite set. Then the parametrizing set A contains the following $3s - 3$ elements

- g_i, $i = 1, \ldots, s$,
- $g_i \circ g_1$, $i = 2, \ldots, s$,
- $g_i \circ g_2$, $i = 3, \ldots, s$.

The result of Theorem 1.6.2 is best possible if G_0 is generated freely by E_0. Hence, if we want to reduce the number of the elements of the parametrizing set A, we have to consider groups with defining relations.

If follows from (1.28) that g_1 and g_2 have intersecting axes. This assumption was needed in proving that k in (1.30) is negative. If the axes of g_1 and g_2 do not intersect, then $k(g_i)$, $k(g_i \circ g_1)$ and $k(g_i \circ g_2)$ do not determine g_i uniquely. This fact will make trouble in the proof of Lemma 1.6.3.

The uniformization theory constitutes the connecting link between surfaces and principal–circle groups. Surfaces are represented as quotients of a disk or a half–plane by principal–circle groups, and principal–circle groups corresponding to a surface are isomorphic to its fundamental group.

We shall see later that Theorem 1.6.2 gives minimal parametrizing sets for principal–circle groups corresponding to non–orientable or non–compact surfaces. It remains the rather complicated case of the compact and orientable surfaces.

The fundamental group of a compact and orientable surface of genus p is generated by $2p$ elements $\gamma_1, \ldots, \gamma_{2p}$ with the defining relation

$$(\gamma_1\gamma_2^{-1}\gamma_1^{-1}\gamma_2)(\gamma_3\gamma_4^{-1}\gamma_3^{-1}\gamma_4)\ldots(\gamma_{2p-1}\gamma_{2p}^{-1}\gamma_{2p-1}^{-1}\gamma_{2p}) = 1. \tag{1.31}$$

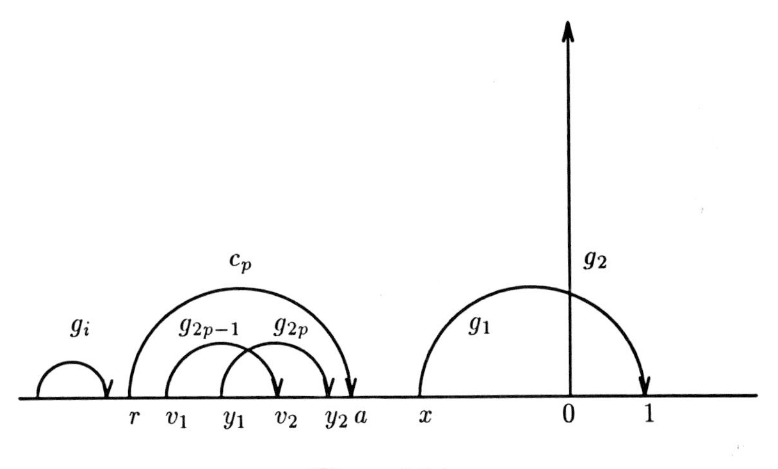

Figure 1.14:

If, in the usual notation, g_k is the Möbius transformation corresponding to γ_k, then $g_k \circ g_i$ corresponds to $\gamma_i\gamma_k$.

For our purposes it is noteworthy in relation (1.31) that it gives a representation for the commutator

$$\gamma_{2p-1}\gamma_{2p}^{-1}\gamma_{2p-1}^{-1}\gamma_{2p}$$

in terms of $\gamma_1, \ldots, \gamma_{2p-2}$.

In order to find a minimal parametrization by multipliers, it suffices to know that principal–circle groups corresponding to compact and orientable surfaces can be characterized as follows.

Consider a set $E = \{g_1, \ldots, g_{2p}\}$ of hyperbolic transformations fixing the upper half–plane U. Suppose that the following conditions are satisfied:

- The transformations $g_i \circ g_1$, $i = 3, \ldots, 2p-2$, and $g_i \circ g_2$, $i = 3, \ldots, 2p$ are non–elliptic.
- The commutator $c_p = [g_{2p-1}, g_{2p}]$ is hyperbolic and it has a given representation in the group generated by $\{g_1, \ldots, g_{2p-2}\}$.
- E is normalized by $a(g_1) = 1$, $a(g_2) = \infty$, $r(g_2) = 0$ and $r(g_1) = x < 0$.
- For any g_i, $i = 3, \ldots, 2p-2$, the cyclic order of the fixed points is given by Figure 1.14.

A set E satisfying the above conditions is said to be of *compact type*. A group G is of *compact type* if it is generated by a set conjugate to E. The number p is called the *genus* of E or G. Hence we allow, for a moment, some ambiguity in the definition of the genus of a group. However, if a compact and orientable surface of genus p is a represented as a quotient U/G, then p is the smallest number satisfying the above definition.

In the following lemma, we give a minimal set of multipliers determining E uniquely.

Lemma 1.6.3 *Let $E = \{g_1, \ldots, g_{2p}\}$ be a set of compact type. Suppose that the classes of the pairs (g_i, g_1), $i = 3, \ldots, 2p-2$, and (g_i, g_2), $i = 3, \ldots, 2p$, are fixed in $\mathcal{P} \cup \mathcal{H}$. Then the following $6p-4$ numbers*

- $k(g_i)$, $i = 1, \ldots, 2p$,
- $k(g_i \circ g_1)$, $i = 2, \ldots, 2p-2$,
- $k(g_i \circ g_2)$, $i = 3, \ldots, 2p$,
- $k(g_{2p-1} \circ g_{2p})$

determine E uniquely. This set is minimal.

Proof. Firstly, give the following $6p-9$ multipliers:

- $k(g_i)$, $i = 1, \ldots, 2p-2$,
- $k(g_i \circ g_1)$, $i = 2, \ldots, 2p-2$,
- $k(g_i \circ g_2)$, $i = 3, \ldots, 2p-2$.

Then, by Lemma 1.6.1, the set $\{g_1, \ldots, g_{2p-2}\}$ and hence also the commutator $c_p = [g_{2p-1}, g_{2p}]$ are determined.

Assume that

- $k(g_{2p-1})$ and $k(g_{2p})$

are given. Then $k(g_{2p-1} \circ g_{2p})$ has (at most) two possible values by Theorem 1.6.2. Hence, by Lemma 1.4.1, also the cross–ratio

$$t_p = (r(g_{2p-1}), r(g_{2p}), a(g_{2p}), a(g_{2p-1}))$$

has at most two possible values. To choose one of these we have to spend one parameter, i.e., we have to suppose that also

- • $k(g_{2p-1} \circ g_{2p})$

is given. Since $(g_{2p-1}, g_{2p}) \in \mathcal{H}$, the pair (g_{2p-1}, g_{2p}) is determined up to conjugation.

Give the multiplier

- $k(g_{2p} \circ g_2)$.

Suppose that (g'_{2p-1}, g'_{2p}) is another candidate for the pair (g_{2p-1}, g_{2p}). Then there exists a Möbius transformation ψ fixing U such that

$$\begin{aligned} g'_{2p-1} &= \psi \circ g_{2p-1} \circ \psi^{-1}, \\ g'_{2p} &= \psi \circ g_{2p} \circ \psi^{-1}. \end{aligned}$$

Since (g_{2p-1}, g_{2p}) and (g'_{2p-1}, g'_{2p}) have the same commutator c_p, we have

$$c_p = \psi \circ c_p \circ \psi^{-1}.$$

Hence ψ fixes the axis of c_p. By the normalization of Figure 1.14, ψ maps the interval $[r, a]$ increasingly onto itself. Hence ψ is hyperbolic or the identity transformation. Since also (g_{2p}, g_2) and (g'_{2p}, g_2) are conjugate, we have by formulae (1.29) and (1.30) two alternatives: Either $\psi = \mathrm{id}$ or

$$\psi(z) = \frac{(ak - r)z - ar(k-1)}{(k-1)z - kr + a}$$

with

$$k = \frac{r(y_1 - a)(y_2 - a)}{a(y_1 - r)(y_2 - r)}.$$

Hence, to distinguish between these two alternatives, we have to spend one more parameter. Suppose, therefore, that also

- • $k(g_{2p-1} \circ g_2)$

is given. To show that then only the case $\psi = \mathrm{id}$ remains, denote $v_1 = r(g_{2p-1})$ and $v_2 = a(g_{2p-1})$. Then, similarly as for g_{2p}, we get an expression for k:

$$k = \frac{r(v_1 - a)(v_2 - a)}{a(v_1 - r)(v_2 - r)}.$$

Hence

$$\frac{v_1}{v_1 - r} \cdot \frac{y_1 - r}{y_1 - a} = \frac{y_2 - a}{y_2 - r} \cdot \frac{v_2 - r}{v_2 - a}.$$

But this is impossible since $(v_1, y_1, a, r) > 1$ and $(y_2, v_2, a, r) < 1$. Hence $\psi = \mathrm{id}$.

We have seen that $6p - 4$ multipliers determine E uniquely. If we drop the multipliers $k(g_{2p-1} \circ g_{2p})$ and $k(g_{2p-1} \circ g_2)$ and keep the remaining $6p - 6$ multipliers fixed, then the set $\{g_1, \ldots, g_{2p-2}\}$ is fixed but the pair (g_{2p-1}, g_{2p}) has at most four alternatives. The remaining $6p - 6$ multipliers are all essential: If one of them is dropped, then there are uncountably infinite set of alternatives with the same $6p - 7$ remaining parameters.

Suppose that the essential $6p-6$ multipliers are given. Then fixing of the value of $k(g_{2p-1} \circ g_{2p})$ reduces the number of the alternatives from four to two and finally $k(g_{2p-1} \circ g_2)$ makes the choice between the remaining two cases. □

In the next sections, we shall investigate the geometric meaning of the "inessential" parameters $k(g_{2p-1} \circ g_{2p})$ and $k(g_{2p-1} \circ g_2)$.

Theorem 1.6.4 *Suppose that the reflection group G_0 is generated by a set $E_0 = \{g_1, \ldots, g_{2p}\}$ of compact type. Then the following $6p-4$ elements*

- g_i, $i = 1, \ldots, 2p$,
- $g_i \circ g_1$, $i = 2, \ldots, 2p-2$,
- $g_i \circ g_2$, $i = 3, \ldots, 2p$,
- $g_{2p-1} \circ g_{2p}$

constitute a minimal parametrizing set A of $J(G_0)$.

Proof. Since the pairs (g_i, g_1), $i = 3, \ldots, 2p-2$, and (g_i, g_2), $i = 3, \ldots, 2p$, are in $\mathcal{P} \cup \mathcal{H}$, Theorem 1.4.2 and Lemma 1.6.3 can be applied, and the assertions follow similarly as in the proof of Theorem 1.6.2. □

Theorem 1.6.2 gives a parametrizing set for $J(G_0)$ containing $6p-3$ elements. Hence the relation of G_0 derived from (1.31) reduces the number of elements of a minimal parametrizing set only by one. In the next sections we shall show that it is possible to parametrize $J(G_0)$ by $6p-6$ numbers if two of them are not multipliers but functions of multipliers.

It is open question whether all minimal parametrizing sets of $J(G_0)$ contain (at least) $6p-4$ elements.

1.7 Orthogonal decompositions and twist parameters

In this section, we derive a parametrization by $6p-6$ parameters for compact type groups of genus p. To that end, we continue the study of the commutator $c = [g, h]$ of two hyperbolic transformations.

Consider two Möbius transformations Φ_1 and Φ_2 fixing e.g. the upper half-plane U, and let L be a non-Euclidean line in U. We say that $h = \Phi_2 \circ \Phi_1$ is an *orthogonal decomposition with respect to L* if either

- Φ_1 and Φ_2 are hyperbolic, their axes are perpendicular and $ax(\Phi_1) = L$

or

- $\Phi_1 = \mathrm{id}$, Φ_2 is hyperbolic and $ax(\Phi_2)$ and L are perpendicular.

Before considering the existence and uniqueness of orthogonal decompositions we prove the following lemma:

Lemma 1.7.1 *Let $h(z) = (az+b)/(cz+d)$, $ad-bc=1$, be a hyperbolic transformation fixing U. Then $abcd \neq 0$ if and only if $h(0) \neq 0, \infty$ and $h(\infty) \neq 0, \infty$. In this case, we may choose $a>0$, $b>0$, $c>0$ and $d>0$ if and only if $r(h) < 0 < a(h)$.*

Proof. Since $h(0) = b/d$ and $h(\infty) = a/c$, the first assertion holds.

Suppose, that $abcd \neq 0$. The fixed points $x_1 = a(h)$ and $x_2 = r(h)$ satisfy the equation

$$cx^2 - (a-d)x - b = 0. \tag{1.32}$$

Since $x_1x_2 = -b/c$, we have $x_1x_2 < 0$ if and only if $bc > 0$.

Suppose that $r(h) < 0 < a(h)$. Then $h(0) = b/d > 0$ and $bc > 0$. Choose $b > 0$. Then $c > 0$ and $d > 0$. From $ad - bc = 1$ it then follows that $a > 0$.

Suppose conversely that $a > 0$, $b > 0$, $c > 0$ and $d > 0$. Then $a(h)r(h) < 0$. Since $h(0) = b/d > 0$, we have $r(h) < 0 < a(h)$. □

Theorem 1.7.2 *Let L be a non-Euclidean line in U and let h be a hyperbolic transformation fixing U. If $ax(h)$ and L intersect, then there exists a unique orthogonal decomposition $h = \Phi_2 \circ \Phi_1$ with respect to L.*

Proof. We may suppose without loss of generality that L is the positive imaginary axis and $a(h) > 0$. Since $ax(h)$ and L intersect, we have $r(h) < 0$. Let $h(z) = (az+b)/(cz+d)$, $ad-bc=1$. By Lemma 1.7.1, we may suppose that $a>0$, $b>0$, $c>0$ and $d>0$.

Let $\Phi_2 \circ \Phi_1$ be an orthogonal decomposition with respect to L. Let $x > 0$ and $-x$ be the fixed points of Φ_2. Then by formulae (1.4)–(1.6),

$$\begin{aligned} \Phi_1(z) &= k'z, \quad k' > 0, \\ \Phi_2(z) &= \frac{(k+1)xz + x^2(k-1)}{(k-1)z + (k+1)x}, \quad k > 0. \end{aligned}$$

We have

$$\Phi_2(\Phi_1(z)) = \frac{(k+1)k'xz + x^2(k-1)}{(k-1)k'z + (k+1)x} = \frac{az+b}{cz+d}$$

if and only if

$$\begin{aligned} a &= \frac{(k+1)k'}{2\sqrt{kk'}}, \quad b = \frac{x(k-1)}{2\sqrt{kk'}}, \\ c &= \frac{(k-1)k'}{2x\sqrt{kk'}}, \quad d = \frac{k+1}{2\sqrt{kk'}}, \end{aligned}$$

i.e., if and only if

$$k' = \frac{a}{d}, \quad x = \sqrt{\frac{ab}{cd}}, \quad k = \frac{\sqrt{ad}+\sqrt{bc}}{\sqrt{ad}-\sqrt{bc}}. \;\square$$

Theorem 1.7.3 *Let (Φ_1, Φ_2) be a principal–circle pair with intersecting axes, i.e., $(\Phi_1, \Phi_2) \in \mathcal{H}_R$. Then*

$$h = \Phi_2 \circ \Phi_1 \tag{1.33}$$

is an orthogonal decomposition if and only if

$$f(h) = \frac{1}{2} f(\Phi_1) f(\Phi_2). \tag{1.34}$$

Proof. By Figure 1.1, (1.33) is orthogonal if and only if

$$t = (r(\Phi_1), r(\Phi_2), a(\Phi_2), a(\Phi_1)) = \frac{1}{2}.$$

Since $(\Phi_1, \Phi_2) \in \mathcal{H}_R$, this occurs if and only if

$$f(h) = \frac{1}{2}[f(k(\Phi_1)k(\Phi_2)) + f(k(\Phi_1)/k(\Phi_2))].$$

The assertion follows by simple calculation. □

Since $f(\mathrm{id}) = 2$, the formula (1.34) holds for all orthogonal decompositions $h = \Phi_2 \circ \Phi_1$.

Let L be an *oriented* non–Euclidean line in U, i.e., denote one of the ideal end points of L by $r(L)$ and the other by $a(L)$ and let the positive direction on L be defined by $r(L) \to a(L)$. The axis of a hyperbolic transformation h has a natural orientation given by $r(h)$ and $a(h)$. Then h translates the points of $ax(h)$ towards $a(h)$, i.e., to the positive direction.

Suppose that L and $ax(h)$ intersect. Let $\alpha(L, h)$ be the angle between L and $ax(h)$ determined by the positive orientations of L and $ax(h)$. In Figure 1.15, we have the same normalization as in the proof of Theorem 1.7.2.

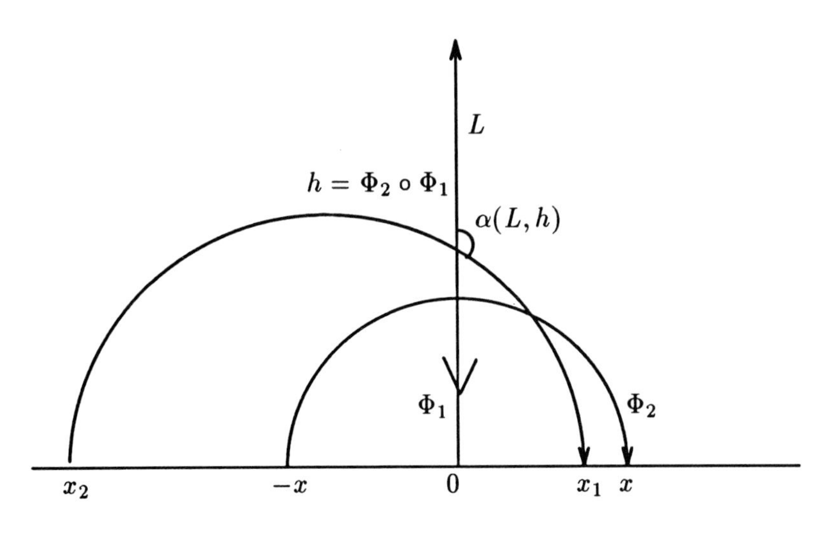

Figure 1.15:

Theorem 1.7.4 *Let $h = \Phi_2 \circ \Phi_1$ be an orthogonal decomposition with respect to L. Then $k(h) \geq k(\Phi_2)$. Moreover*

- $\alpha(L,h) = \pi/2 \Longleftrightarrow \Phi_1 = \mathrm{id} \Longleftrightarrow k(h) = k(\Phi_2)$,
- $\alpha(L,h) < \pi/2 \Longleftrightarrow a(\Phi_1) = a(L)$,
- $\alpha(L,h) > \pi/2 \Longleftrightarrow a(\Phi_1) = r(L)$.

Proof. Since $f(\Phi_1) \geq 2$, we have by (1.34) $f(h) \geq f(\Phi_2)$. Then $k(h) \geq k(\Phi_2)$ because the function $k \mapsto f(k)$ is strictly increasing for $k \geq 1$. The equality holds if and only if $f(\Phi_1) = 2$, i.e., $\Phi_1 = \mathrm{id}$. In this case, $h = \Phi_2$ and $\alpha(L,h) = \pi/2$.

Normalize by $a(L) = \infty$, $r(L) = 0$, $x_1 = a(h) > 0$ and $x_2 = r(h) < 0$ (Figure 1.15). Let $h(z) = (az+b)/(cz+d)$, $ad - bc = 1$ with $a > 0$, $b > 0$, $c > 0$ and $d > 0$ (cf. Lemma 1.7.1). Then by (1.32) $x_1 + x_2 = (a-d)/c$. On the other hand, by the proof of Theorem 1.7.2,

$$\Phi_1(z) = \frac{a}{d} z.$$

Hence

$$\alpha(L,h) < \pi/2 \Longleftrightarrow x_1 + x_2 > 0 \Longleftrightarrow a > d \Longleftrightarrow a(\Phi_1) = a(L),$$
$$\alpha(L,h) > \pi/2 \Longleftrightarrow x_1 + x_2 < 0 \Longleftrightarrow a < d \Longleftrightarrow a(\Phi_1) = r(L). \square$$

Consider a pair $(g,h) \in \mathcal{H}_R$ whose commutator

$$c = [g,h] = h \circ g^{-1} \circ h^{-1} \circ g$$

is hyperbolic. Let $h = \Phi_2 \circ \Phi_1$ be the orthogonal decomposition with respect to $L = ax(g)$. Then $\Phi_1^{\pm 1}$ and $g^{\pm 1}$ commute, and it follows that

$$c = [g,h] = [g,\Phi_2].$$

Hence only the component Φ_2 of h whose axis is perpendicular to $ax(g)$ contributes to the commutator of g and h. By Theorem 1.7.4,

$$k(h) \geq k(\Phi_2).$$

Moreover, by Theorem 1.5.3 and its Corollary 1.5.4,

$$f(\Phi_2^2) = \frac{4(f(c)+2)}{f(g^2)-2} + 2.$$

Hence $k(c)$ and $k(g)$ determine $k(\Phi_2)$ uniquely.

Let $k_1 = k(g)$ and $k_4 = k(c)$. Define $k_0 > 1$ by the formula

$$f(k_0^2) = \frac{4(f(k_4)+2)}{f(k_1^2)-2} + 2.$$

On the other hand, define

$$\epsilon(g,h) = \begin{cases} 0 & \text{if } \Phi_1 = \mathrm{id}, \\ 1 & \text{if } a(\Phi_1) = a(g), \\ -1 & \text{if } a(\Phi_1) = r(g). \end{cases}$$

Hence $k_0 \leq k(h)$ and $k_0 = k(h)$ if and only if $\epsilon(g,h) = 0$.

Consider the number

$$m = \left[\frac{k(h)}{k_0}\right]^{\epsilon(g,h)}.$$

Let $\alpha(g,h)$ be the angle determined by the positive directions of the axes of g and h. Then, by Theorem 1.7.4,

- $\alpha(g,h) = \pi/2 \iff \epsilon(g,h) = 0 \iff m = 1,$
- $\alpha(g,h) < \pi/2 \iff \epsilon(g,h) = 1 \iff m > 1,$
- $\alpha(g,h) > \pi/2 \iff \epsilon(g,h) = -1 \iff m < 1.$

Theorem 1.7.5 *Let $(g,h) \in \mathcal{H}_R$ be a pair with hyperbolic $c = [g,h]$. If the multipliers $k_1 = k(g)$ and $k_4 = k(c)$ and the number*

$$m = m(g,h) = \left[\frac{k(h)}{k_0}\right]^{\epsilon(g,h)}$$

are known, then the pair (g,h) is determined up to conjugation.

Proof. Since k_1 and k_4 determine k_0, the number m determines then both $k_2 = k(h)$ and $\epsilon(g,h)$. By Corollary 1.5.4 of Theorem 1.5.3, the multipliers k_1, k_2 and k_4 determine the acute angle between the axes of g and h. On the other hand, $\epsilon(g,h)$ tells which one of the adjacent angles formed by the axes of g and h is acute. Hence $t = (r(g), r(h), a(h), a(g))$ is uniquely determined and the assertion follows. □

Consider a set $E = \{g_1, \ldots, g_{2p}\}$ of compact type. The first $6p-9$ multipliers given in the proof of Lemma 1.6.3 determine the commutator

$$c_p = [g_{2p-1}, g_{2p}].$$

Then by Theorem 1.7.5, the numbers $k(g_{2p-1})$ and $m(g_{2p-1}, g_{2p})$ determine the pair (g_{2p-1}, g_{2p}) up to conjugation. The only freedom left is conjugation by hyperbolic transformations fixing the axis of c_p. Geometrically, the "handle" determined by (g_{2p-1}, g_{2p}) is then "rotated around" the commutator c_p. This rotation can be determined by one real parameter, so called *twist parameter*, as follows.

Let L be a non–Euclidean line in U and let $\Sigma(L)$ denote the group generated by the hyperbolic transformations $\psi : U \to U$ with $ax(\psi) = L$. A family $\mathcal{F}$ of pairs (g,h) of hyperbolic transformations fixing U is *L–invariant* if the following conditions are satisfied:

- If $(g,h) \in \mathcal{F}$ and $\psi \in \Sigma(L)$ then $\psi(g,h)\psi^{-1} = (\psi \circ g \circ \psi^{-1}, \psi \circ h \circ \psi^{-1}) \in \mathcal{F}$.

- If $(g_1,h_1) \in \mathcal{F}$ and $(g_2,h_2) \in \mathcal{F}$, then there exists a uniquely determined $\psi \in \Sigma(L)$ for which $(g_2,h_2) = \psi(g_1,h_1)\psi^{-1}$.

Fix $(g_0,h_0) \in \mathcal{F}$ and suppose that $\mathcal{F}$ is L–invariant. The mapping

$$(g,h) \mapsto \psi \quad \text{if} \quad (g,h) = \psi(g_0,h_0)\psi^{-1} \tag{1.35}$$

is a bijection $\mathcal{F} \to \Sigma(L)$. Choose an orientation $r(L) \to a(L)$ on L. For $(g,h) = \psi(g_0,h_0)\psi^{-1}$, call the number

$$\omega((g,h),(g_0,h_0)) = \begin{cases} \log k(\psi) & \text{if} \quad a(\psi) = a(L), \\ -\log k(\psi) & \text{if} \quad a(\psi) = r(L), \end{cases} \tag{1.36}$$

the *twist parameter* of $(g,h) \in \mathcal{F}$ with (g_0,h_0) as the origin. Hence, by (1.35), every $(g_0,h_0) \in \mathcal{F}$ defines a bijection

$$\omega(\bullet,(g_0,h_0)) : \mathcal{F} \to \mathbf{R}.$$

We can now choose the right element from an L-invariant family $\mathcal{F}$ as soon as we fix an element $(g_0,h_0) \in \mathcal{F}$ and give the twist parameter of the required element.

Let $c : U \to U$ be a given hyperbolic transformation with $r(c) < a(c) < 0$. Consider all pairs $(g,h) \in \mathcal{H}_R$ for which

$$r(c) < r(g) < r(h) < a(g) < a(h) < a(c)$$

and

$$c = [g,h].$$

Then, by Theorem 1.7.5, the pairs (g,h) with given values of

$$k = k(h)$$

and

$$m = m(g,h)$$

constitute an $ax(c)$-invariant family $\mathcal{F}$. Without loss of generality we may consider e.g. this family in the following theorems.

To parametrize $\mathcal{F}$, it suffices to define the origin (g_0,h_0) of the twist parameter (1.36) in an invariant way. Let g_2 be a given hyperbolic transformation with $r(g_2) = 0$ and $a(g_2) = \infty$ (see Figure 1.16).

Theorem 1.7.6 *Suppose that* $(g_2,c) \in \operatorname{Int}\mathcal{P}$. *Then* $(h,g_2) \in \mathcal{P}$ *for all* $(g,h) \in \mathcal{F}$ *and the function*

$$(g,h) \mapsto k(h \circ g_2) \tag{1.37}$$

has its absolute minimum at a uniquely determined point $(g_0,h_0) \in \mathcal{F}$.

Proof. By Lemma 1.5.2, $(h, g^{-1} \circ h^{-1} \circ g) \in \operatorname{Int}\mathcal{P}$. Then by Lemma 1.5.9, $(g^{-1} \circ h \circ g \circ h^{-1}, h) = (c^{-1}, h) \in \operatorname{Int}\mathcal{P}$ and hence also $(h, c^{-1}) \in \operatorname{Int}\mathcal{P}$. Since $(g_2, c) \in \operatorname{Int}\mathcal{P}$, we have $(h, g_2) \in \operatorname{Int}\mathcal{P}$ by Theorem 1.5.11.

Fix $(g_1,h_1) \in \mathcal{F}$. Then for any $(g,h) \in \mathcal{F}$ there exists a uniquely determined $\varphi \in \Sigma(ax(c))$ such that

$$h = \varphi \circ h_1 \circ \varphi^{-1}.$$

Denote $r = r(c)$ and $a = a(c)$. Then by (1.29)

$$\varphi(z) = \frac{(a\xi - r)z - ar(\xi - 1)}{(\xi - 1)z - \xi r + a}, \quad \xi > 0.$$

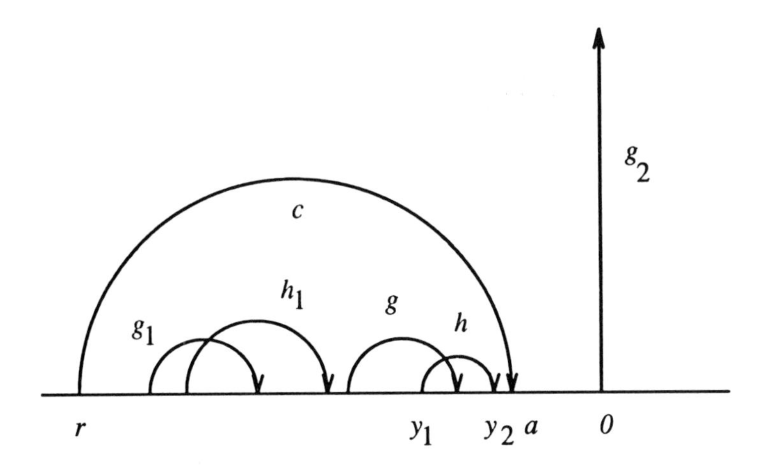

Figure 1.16:

The cross–ratio t associated with the pair (h, g_2) depends on ξ:

$$t(\xi) = (r(h), r(g_2), a(g_2), a(h)) = \frac{\varphi(a(h_1))}{\varphi(a(h_1)) - \varphi(r(h_1))}. \tag{1.38}$$

If $\xi \to \infty$ or $\xi \to 0+$, then $r < \varphi(a(h_1)) < a < 0$ whereas $\varphi(a(h_1)) - \varphi(r(h_1)) \to 0+$. Hence

$$t(\xi) \to -\infty \qquad \text{as} \quad \xi \to \infty \quad \text{or} \quad \xi \to 0+.$$

Since, by the definition of the class $\mathcal{P}$,

$$f(h \circ g_2) = -t(\xi)[f(k(h_1)k(g_2)) - f(k(h_1)/k(g_2))] - f(k(h_1)/k(g_2)),$$

we have

$$K(\xi) = k(h \circ g_2) \to \infty \tag{1.39}$$

as $\xi \to \infty$ or $\xi \to 0+$. Moreover, $K(\xi)$ is a continuous function of $\xi > 0$.

Let $(g', h') \in \mathcal{F}$ be the pair with the parameter $\xi' > 0$. Suppose that $K(\xi) = K(\xi')$, i.e., $k(h \circ g_2) = k(h' \circ g_2)$. Since $k(h) = k(h')$ and the pairs (h, g_2) and (h', g_2) are in $\mathcal{P}$, they are conjugate (Lemma 1.4.1), i.e., there exists a hyperbolic transformation σ fixing 0 and ∞ such that

$$\sigma \circ h \circ \sigma^{-1} = h'. \tag{1.40}$$

On the other hand, since (g, h) and (g', h') are in $\mathcal{F}$, there exists $\psi \in \Sigma(ax(c))$, such that

$$\psi \circ h \circ \psi^{-1} = h'.$$

Let, by (1.29),

$$\psi(z) = \frac{(ak-r)z - ar(k-1)}{(k-1)z - kr + a}, \quad k > 0.$$

Denote $y_1 = r(h)$ and $y_2 = a(h)$. Then

$$\frac{\psi(y_1)}{\psi(y_2)} = \frac{\sigma(y_1)}{\sigma(y_2)} = \frac{y_1}{y_2},$$

and we have by (1.24)

$$k = \frac{r(y_1 - a)(y_2 - a)}{a(y_1 - r)(y_2 - r)}. \tag{1.41}$$

Hence $K(\xi) = k(h \circ g_2)$ attains a given value at most twice. Then by (1.39), the function K is convex and there exists a uniquely determined pair $(g_0, h_0) \in \mathcal{F}$ such that the function (1.37) has its absolute minimum at (g_0, h_0). □

The minimum of $K(\xi) = k(h \circ g_2)$ can be characterized as the unique value attained by $K(\xi)$ once only. This in turn occurs if and only if $k = 1$ in (1.41). Hence Theorem 1.7.6 has the following

Corollary 1.7.7 *Under the hypotheses of Theorem 1.7.6, the function (1.37) has its absolute minimum at (g_0, h_0) if and only if*

$$\frac{r(r(h_0) - a)(a(h_0) - a)}{a(r(h_0) - r)(a(h_0) - r)} = 1. \;\square$$

By considering hyperbolic geometry in U, Theorem 1.7.6 can be generalized as follows. Let d_U denote the hyperbolic distance in U defined by the line element

$$ds = \frac{|dz|}{\operatorname{Im} z}.$$

Denote by $d_U(L_1, L_2)$ the shortest distance between two non–Euclidean lines L_1 and L_2.

Theorem 1.7.8 *Let $\mathcal{F}$ be an $ax(c)$–invariant family and g_2 a hyperbolic transformation satisfying the normalization of Figure 1.16. Then the following holds:*

- *there exists a unique $(g_0, h_0) \in \mathcal{F}$ such that*

$$d_U(ax(h_0), ax(g_2)) \le d_U(ax(h), ax(g_2))$$

for all $(g, h) \in \mathcal{F}$,

- *if* $(h_0, g_2) \in \mathcal{P}$, *then* $(h, g_2) \in \mathcal{P}$ *for all* $(g,h) \in \mathcal{F}$,
- *if* $(h_0, g_2) \in \mathcal{P}$, *then the function* $(h,g) \mapsto k(h \circ g_2)$ *has its absolute minimum in* $\mathcal{F}$ *at* (g_0, h_0).

Proof. For $(g,h) \in \mathcal{F}$ and $(g', h') \in \mathcal{F}$ we have

$$d_U(ax(h), ax(g_2)) = d_U(ax(h'), ax(g_2))$$

if and only if there exists a hyperbolic σ fixing 0 and ∞ such that $h' = \sigma \circ h \circ \sigma^{-1}$, i.e., (1.40) holds. The existence of (g_0, h_0) follows now similarly as in the proof of Theorem 1.7.6 by replacing $K(\xi)$ by the distance $d_U(ax(h), ax(g_2))$. Also at this minimum point we have $k = 1$ in (1.41).

For $(g,h) \in \mathcal{F}$ let $t = t(\xi)$ be defined by (1.38). Then by simple calculation,

$$d_U(ax(h), ax(g_2)) = \log(1 - 2t + 2\sqrt{t^2 - t})$$

which, as a function of t, is decreasing for the values $t < 0$. Hence $t(\xi)$ attains its maximum $t_0 = t(\xi_0)$ for the pair (g_0, h_0). Then $t(\xi) \leq t(\xi_0)$ and $(h, g_2) \in \mathcal{P}$ if $(h_0, g_2) \in \mathcal{P}$.

The last assertion follows now by Theorem 1.7.6 and its Corollary 1.7.7.

□

Corollary 1.7.9 *For the family* $\mathcal{F}$ *in Theorem 1.7.8, choose* (g_0, h_0) *minimizing*

$$d_U(ax(h), ax(g_2))$$

as the origin of the twist parameter (1.36). If $(h_0, g_2) \in \mathcal{P}$, *then the multiplier* $k(h \circ g_2)$ *is a convex function of the twist parameter*

$$x = \omega((g,h),(g_0,h_0))$$

and as a function of x, $k(h \circ g_2)$ *attains its minimum in* $\mathbf{R}$ *at* $x = 0$.

Proof. The convexity of the function $x \mapsto k(h \circ g_2)$ follows from the convexity of the function (1.39). By Theorem 1.7.8, (g_0, h_0) minimizes also $k(h \circ g_2)$.

□

Suppose that $(g_2, c) \in \operatorname{Int} \mathcal{P}$. Then by Theorem 1.7.6 $(h_0, g_2) \in \mathcal{P}$. For $(g,h) \in \mathcal{F}$ define

$$\epsilon(h) = \begin{cases} 1 & \text{if} \quad \omega((g,h),(g_0,h_0)) \geq 0, \\ -1 & \text{if} \quad \omega((g,h),(g_0,h_0)) < 0. \end{cases}$$

Then we have, by the above corollary, the following result:

Theorem 1.7.10 *The function*

$$(g,h) \mapsto \log \left[\frac{k(h \circ g_2)}{k(h_0 \circ g_2)}\right]^{\epsilon(h)}$$

is a bijection $\mathcal{F} \to \mathbf{R}$. □

In order to apply the preceding considerations to the parametrization of compact type groups, fix the genus p and consider sets $E = \{g_1, \ldots, g_{2p}\}$ of compact type for which $(g_2, c_p) \in \operatorname{Int} \mathcal{P}$. Give first the following $6p-8$ multipliers:

- $k(g_i)$, $i = 1, \ldots, 2p-1$,
- $k(g_i \circ g_1)$, $i = 2, \ldots, 2p-2$,
- $k(g_i \circ g_2)$, $i = 3, \ldots, 2p-2$.

Then the set $\{g_1, \ldots, g_{2p-2}\}$ and hence also $c_p = [g_{2p-1}, g_{2p}]$ are determined.

Consider all sets E of genus p with these $6p-8$ parameters fixed. Let $g_{2p} = \Phi_2 \circ \Phi_1$ be the orthogonal decomposition with respect to $ax(g_{2p-1})$. Then the numbers $k(c_p)$ and $k(g_{2p-1})$ determine $k(\Phi_2)$ and we have $k(g_{2p}) \geq k(\Phi_2)$. Hence, if we consider all possible pairs $(g,h) = (g_{2p-1}, g_{2p})$, there exists a minimum

$$k_0 = \min_E k(g_{2p}).$$

Fix the number

- $m_1 = \left[\frac{k(g_{2p})}{k_0}\right]^{\epsilon(g_{2p-1}, g_{2p})}$.

Then the $ax(c_p)$–invariant family $\mathcal{F}$ containing all pairs $(g,h) = (g_{2p-1}, g_{2p})$ with the given values of $k(g_{2p-1})$ and $m(g_{2p-1}, g_{2p})$ is determined. We may apply Theorem 1.7.6 to $\mathcal{F}$. Let

$$k_* = \min_E k(g_{2p} \circ g_2).$$

Then, finally, the number

- $m_2 = \left[\frac{k(g_{2p} \circ g_2)}{k_*}\right]^{\epsilon(g_{2p})}$

determines the pair (g_{2p-1}, g_{2p}).

Fix a compact type set $E_0 = \{h_1, \ldots, h_{2p}\}$ of genus p. Denote $\gamma_p = [h_{2p-1}, h_{2p}]$ and suppose that $(h_2, \gamma_p) \in \operatorname{Int} \mathcal{P}$. Let G_0 be the principal–circle group generated by E_0. Define in $J(G_0)$ an equivalence relation $\sim$ by setting

$(j_1, G_1) \sim (j_2, G_2)$ if $j_2 \circ j_1^{-1} : G_1 \to G_2$ is induced by a Möbius transformation. Then every equivalence class $[j, G]$ contains a unique representative (j, G) such that the set $j(E_0) = \{g_1, \ldots, g_{2p}\}$ with $g_i = j(h_i)$, $i = 1, \ldots, 2p$, is of compact type. Moreover $(g_2, c_p) \in \operatorname{Int} \mathcal{P}$ by Theorem 1.4.2. We have proved the following result:

Theorem 1.7.11 *The mapping $J(G_0)/\sim \to \mathbf{R}^{6p-6}$ defined by*

- $x_i = \log k(g_i)$, $i = 1, \ldots, 2p-1$,
- $x_{2p-1+i} = \log k(g_i \circ g_1)$, $i = 2, \ldots, 2p-2$,
- $x_{4p-4+i} = \log k(g_i \circ g_2)$, $i = 3, \ldots, 2p-2$,
- $x_{6p-7} = \log \left[\frac{k(g_{2p})}{k_0}\right]^{\epsilon(g_{2p-1}, g_{2p})}$,
- $x_{6p-6} = \log \left[\frac{k(g_{2p} \circ g_2)}{k_*}\right]^{\epsilon(g_{2p})}$

is injective. □

We shall see later that the coordinates x_i, $i = 1, \ldots, 6p-6$, have rather simple geometrical interpretations on a compact and orientable surface of genus p.

Chapter 2

Quasiconformal mappings

2.1 Introduction to Chapter 2

It is our ultimate goal to give parameters which define Riemann surfaces up to conformal isomorphisms. This is the famous moduli problem.

The first solution was based on the theory of quasiconformal mappings. They give a way to deform one conformal type of Riemann surfaces to another conformal type. In this Chapter we will review the necessary part of the theory of quasiconformal mappings. Everything here belongs to the classical foundations of this theory. Consequently, we will, in this Chapter, give only a quick review of quasiconformal mappings omitting all the proofs. We try to keep the presentations as readable as possible giving exact references for all the omitted proofs.

Our main references for quasiconformal mappings are the monograph of Lars Ahlfors [6] and that of Olli Lehto and K. I. Virtanen [59].

We will start with describing conformal invariants of certain simple plain domains. These invariants will then be used to define quasiconformal mappings and to study their properties. In the subsequent chapters quasiconformal mappings will be, in turn, applied to find conformal invariants which determine the isomorphism class of a Riemann surface.

2.2 Conformal invariants

Let Q be a simply connected closed set in the extended complex plane $\hat{\mathbf{C}}$. Assume that the boundary of Q is a Jordan curve. Let p_1, p_2, p_3 and p_4 be disjoint points of ∂Q whose cyclic order agrees with the positive orientation of ∂Q.

Definition 2.2.1 *The domain Q together with the disjoint points p_1, p_2, p_3 and p_4 on ∂Q is* a quadrilateral $Q = Q(p_1, p_2, p_3, p_4)$. *For $i = 1, 2$, let*

$Q_i(p_1^i, p_2^i, p_3^i, p_4^i)$ *be quadrilaterals. A mapping* $f : Q_1 \to Q_2$ *is a* homeomorphisms of quadrilaterals *if* f *is a homeomorphic mapping of the closed set* Q_1 *onto the closed set* Q_2 *satisfying* $f(p_j^1) = f(p_j^2)$ *for all* $j = 1, 2, 3, 4$. *A homeomorphism of quadrilaterals* Q_1 *and* Q_2 *is* an isomorphism *if it is holomorphic in the interior of* Q_1.

This is the standard definition, see e.g. [59, Page 15].

Lemma 2.2.1 *Let* $Q(p_1, p_2, p_3, p_4)$ *be a quadrilateral. The exists a constant* $M = M(Q) \geq 1$ *such that* Q *is isomorphic to the quadrilateral*

$$N(0, M, M + i, i), \quad N = \{z \in \mathbf{C} | 0 \leq Re\, z \leq M, 0 \leq Im\, z \leq i\}.$$

Proof. [59, Page 15].

Definition 2.2.2 *The number* M *of Lemma 2.2.1 is called* the modulus of *the quadrilateral* Q.

It is obvious that two quadrilaterals are isomorphic if and only if they have the same modulus. We conclude that the modulus is a conformal invariant that determines the isomorphism class of a quadrilateral.

Ring domains form another class of domains that can be used to define quasiconformal mappings.

Definition 2.2.3 *An open subset* B *of the extended complex plane* $\hat{\mathbf{C}}$ *which is homeomorphic to the annulus* $\{z | 1 < |z| < 2\}$ *is called* a ring domain. *A ring domain* B *is* degenerate *if one of its boundary components is a point. Otherwise* B *is* non-degenerate.

We also call ring domain *a doubly connected domain.*

It can be proved with the help of harmonic measures that a ring domain $B \subset \hat{\mathbf{C}}$ can be mapped conformally onto an annulus $\{z | 0 \leq r_1 < |z| < r_2 \leq \infty\}$, see eg. [6, 6:5.1]. If B is non-degenerate, then $0 < r_1 < r_2 < \infty$. In this case the ratio r_2/r_1 is the same for all annuli conformally equivalent to B.

Definition 2.2.4 *The number*

$$M(B) = \log \frac{r_2}{r_1} > 0$$

is the modulus of *a non-degenerate ring domain* B *which is conformally equivalent to* $\{z | 0 < r_1 < |z| < r_2 < \infty\}$. *For a degenerate* B, *set* $M(A) = \infty$.

- The modulus of non–degenerate ring domains is always finite and positive.
- The modulus is monotonic: If $B' \subset B$ is a ring domain separating the boundary components of B, then $M(B') \leq M(B)$. [59, I.6.6.].
- The modulus is a conformal invariant: If f is a Möbius transformation, then $M(f(B)) = M(B)$.

2.3 Definitions for quasiconformal mappings

It is interesting to observe that the modulus of a ring domain or that of a quadrilateral behave in the same way under deformations. The following result can be found for instance in the monograph of Lehto and Virtanen ([59, Theorem I.7.2, page 39]).

Lemma 2.3.1 *Let $f : A \to f(A) \subset \hat{\mathbf{C}}$ be a sense-preserving homeomorphism of an open subset of the extended complex plane. The following conditions are equivalent:*

- *The exists a constant $K \geq 1$ such that for all quadrilaterals*

$$Q(p_1, p_2, p_3, p_4)$$

  in A we have

$$\frac{M(f(Q))}{K} \leq M(Q) \leq KM(f(Q)). \tag{2.1}$$

- *The exists a constant $K \geq 1$ such that for all ring domains B in A we have*

$$\frac{M(f(B))}{K} \leq M(B) \leq KM(f(B)). \tag{2.2}$$

Definition 2.3.1 *Let $A \subset \hat{\mathbf{C}}$ be an open set. Let $K \geq 1$. A sense preserving homeomorphism $f : A \to f(A)$ is K–*quasiconformal *if the equivalent conditions (2.1) and (2.2) are fulfilled for each quadrilateral in A and for each ring domain in A. A mapping f is called* quasiconformal *if it is K-quasiconformal for some constant K, $K \geq 1$. The smallest constant K for which a quasiconformal mapping f is K-quasiconformal is called* the maximal dilatation of f. *We use also the notation $K(f)$ for the maximal dilatation of a quasiconformal mapping f.*

This definition for quasiconformal mappings is, in some sense, global. We can also define quasiconformality in terms of local conditions.

Let A be an open subset of the complex plane $\mathbf{C}$. Let $f : A \to f(A) \subset \mathbf{C}$ be a sense–preserving homeomorphism.

Definition 2.3.2 The circular distortion of the mapping f at a point $z \in A$ *is the number*

$$H(z) = \limsup_{r \to 0+} \frac{max_{\varphi}|f(z + re^{i}\varphi) - f(z)|}{min_{\varphi}|f(z + re^{i}\varphi) - f(z)|}. \tag{2.3}$$

Lemma 2.3.2 *A sense-preserving homeomorphism* $f : A \to f(A) \subset \mathbf{C}$ *is* K-quasiconformal, $K \geq 1$, *if the circular distortion* $H(z)$ *of* f *is bounded in* A *and* $H(z) \leq K$ *almost everywhere in* A.

Proof. [59, Pages 177–178].

The following properties of quasiconformal mappings follow from the definition:

- If f_1 and f_2 are quasiconformal, then also $f = f_1 \circ f_2$ is quasiconformal and $K(f) \leq K(f_1)K(f_2)$.
- $K(f) = K(f^{-1})$.
- If f is quasiconformal and g and h are orientation preserving Möbius transformations, then $K(f) = K(g \circ f \circ h)$.

2.4 Complex dilatation

Quasiconformal mappings that are regular enough can easily be characterized analytically. For a *diffeomorphism* $f : A \to f(A)$ define *the complex derivates* setting

$$\partial f = \frac{1}{2}(f_x - if_y), \quad \overline{\partial} f = \frac{1}{2}(f_x + if_y).$$

Here f_x and f_y denote the partial derivates of f with respect to x and to y, $z = x + iy$, respectively. The *derivate of* f *in the direction* α, $0 \leq \alpha \leq 2\pi$, is

$$\partial_\alpha f(z) = \lim_{r \to 0+} \frac{f(z + re^{i\alpha}) - f(z)}{re^{i\alpha}}.$$

Clearly,

$$\partial_\alpha f = \partial f + \overline{\partial} f e^{-2i\alpha}.$$

Therefore,

$$\max_\alpha |\partial_\alpha f(z)| = |\partial f(z)| + |\overline{\partial} f(z)|, \quad \min_\alpha |\partial_\alpha f(z)| = |\partial f(z)| - |\overline{\partial} f(z)|.$$

The Jacobian $J_f = |\partial f|^2 - |\overline{\partial} f|^2$ of *a sense preserving* diffeomorphism f is positive. Therefore, if f is *a sense preserving* diffeomorphism,

$$\min_\alpha |\partial_\alpha f(z)| = |\partial f(z)| - |\overline{\partial} f(z)| > 0.$$

Definition 2.4.1 *The number*

$$D_f(z) = \frac{\max_\alpha |\partial_\alpha f(z)|}{\min_\alpha |\partial_\alpha f(z)|} = \frac{|\partial f(z)| + |\overline{\partial} f(z)|}{|\partial f(z)| - |\overline{\partial} f(z)|}$$

is the dilatation quotient *of the mapping f at a point z.*

Lemma 2.4.1 *A sense preserving diffeomorphism $f : A \to f(A)$ is K-quasiconformal, $K \geq 1$, if and only if*

$$D_f(z) \leq K \tag{2.4}$$

for every $z \in A$.

Proof. [60, Theorem 3.1, page 19].

Definition 2.4.2 *A continuous function $u : A \to \mathbf{R}$ is* absolutely continuous on lines *(ACL) in a domain $A \subset \mathbf{C}$ if for each rectangle $\{x + iy | a \leq x \leq b, c \leq y \leq d\} \subset A$, the function $x \mapsto u(x + iy)$ is of bounded variation on $[a, b]$ for almost all $y \in [c, d]$ and $y \mapsto u(x + iy)$ is of bounded variation on $[c, d]$ for almost all $x \in [a, b]$. A complex valued function f is ACL in A if its real and imaginary parts are ACL in A.*

Lemma 2.4.2 *A quasiconformal mapping is absolutely continuous on lines.*

Proof. [92].

It follows from standard theorems of real analysis that a function f which is ACL in A has finite partial derivates f_x and f_y a.e. in A. We conclude that a quasiconformal mapping has finite partial derivates a.e.

Lemma 2.4.3 *Quasiconformal mappings are differentiable almost everywhere.*

Proof. [33].

We conclude that a quasiconformal mapping f of a domain A is differentiable a. e. and satisfies the condition (2.4) a. e.

Theorem 2.4.4 *A sense preserving homeomorphism $f : A \to f(A) \subset \hat{\mathbf{C}}$ is K-quasiconformal if and only if the following holds:*

- *f in ACL in A.*
- $\max_\alpha |\partial_\alpha f(z)| \leq K \min_\alpha |\partial_\alpha (z)|$ *almost everywhere in A.*

Proof. [60, Theorem I.3.5].

This result leads to the important notion of complex dilatation. Let $f : A \to f(A)$ be a K–quasiconformal mapping and z a point where f is differentiable and $J_f(z) > 0$.

Since $\max_\alpha |\partial_\alpha f| = |\partial f| + |\overline{\partial} f|$, $\min_\alpha |\partial_\alpha f| = |\partial f| - |\overline{\partial} f|$, the condition (2.4) is equivalent to

$$|\overline{\partial} f(z)| \leq \frac{K-1}{K+1} |\partial f(z)|. \tag{2.5}$$

Since $J_f(z) = |\partial f(z)|^2 - |\overline{\partial} f(z)|^2 > 0$, $\partial f(z) \neq 0$, and we can form the quotient

$$\mu_f(z) = \frac{\overline{\partial} f(z)}{\partial f(z)}. \tag{2.6}$$

Definition 2.4.3 *The function μ_f which is defined almost everywhere by the formula (2.6) is* the complex dilatation *of the quasiconformal mapping f.*

- Since f is continuous, μ_f is a Borel measurable function.
- By (2.5), $|\mu_f(z)| \leq k = \frac{K-1}{K+1} < 1$ almost everywhere.

Definition 2.4.4 *A Borel measurable function $\mu : A \to \mathbf{C}$ which satisfies*

$$ess\ sup_z |\mu(z)|| < 1$$

is a Beltrami differential *in the domain A. Let μ be a Beltrami differential in A. The differential equation*

$$\overline{\partial} f = \mu \partial f \tag{2.7}$$

is called a Beltrami equation.

For a holomorphic mapping f, μ_f vanishes identically, and (2.7) reduces to the Cauchy–Riemann equation $\overline{\partial} f = 0$.

Definition 2.4.5 *An ACL function f is said* to have L^p–derivates, *if the partial derivates of f are locally in L^p. Such a function f is called* an L^p–solution of *(2.7) in a domain A if (2.7) holds almost everywhere in A.*

Theorem 2.4.5 *A homeomorphism $f : A \to f(A)$ is K-quasiconformal if and only if f is an L^2-solution of an equation $\overline{\partial} f = \mu\, \partial f$, where μ satisfies condition (2.5) for almost all z.*

Proof. [60, Theorem I.4.1].

Let f and g be quasiconformal mappings of a domain A with complex dilatations μ_f and μ_g, respectively. Direct computation yields the transformation rule

$$\mu_{f \circ g^{-1}}(\zeta) = \frac{\mu_f(z) - \mu_g(z)}{1 - \mu_f(z)\overline{\mu_g(z)}} \left(\frac{\partial g(z)}{|\partial g(z)|}\right)^2, \quad \zeta = g(z), \tag{2.8}$$

which is valid for almost all $z \in A$, and hence for almost all $\zeta \in g(A)$.

The above transformation rule can be easily computed but it has deep consequences. Writing $(\partial g(z)/|\partial g(z)|)^2 = a$, $\mu_g(z) = b$, $\mu_f(z) = \xi$ and $\mu_{f \circ g^{-1}} = \omega$, the formula (2.8) becomes

$$\omega = \frac{a\xi - ab}{1 - \xi\bar{b}}.$$

The mapping $\xi \mapsto \omega$ is a Möbius–transformation mapping the unit disk onto itself. We conclude that *the complex dilatation of $\mu_{f \circ g^{-1}}$ depends holomorphically on the complex dilatation μ_f.* This fact plays a crucial role when defining the complex structure of a Teichmüller space.

The relationship between quasiconformal mappings and their Beltrami differentials or complex dilatations is very close. For our purposes they are simply two views of one object. That follows from the following important result.

Theorem 2.4.6 (Existence Theorem) *Let μ be a measurable function in a domain A. Assume that $\|\mu\|_\infty < 1$. There exists a quasiconformal mapping of A whose complex dilatation agrees with μ almost everywhere in A.*

Proof. [58, p. 136] or [59, p. 191].

This is one of the cornerstones of the classical theory of quasiconformal mappings. Assume that f and g are both μ–quasiconformal mappings of the domain A. Observe that by (2.8) the complex dilatation of the mapping $f \circ g^{-1}$ is then identically 0, i.e., $f \circ g^{-1}$ is then a holomorphic homeomorphism. In this sense solutions to the Beltrami differential equation are also unique. This fact is sometimes referred to as the *uniqueness of quasiconformal mappings* and it will be used later.

Actually even more is true: a quasiconformal mapping depends *holomorphically* on its complex dilatation. In order to clarify this statement, which is originally due to Lars V. Ahlfors and Lipman Bers ([2]) we need to recall the concept of holomorphic maps between Banach spaces.

Let E and F be complex Banach spaces and $U \subset E$ an open set.

Definition 2.4.6 *A function $f : U \to F$ has a* derivative *at a point $x_0 \in U$ if there exists a continuous complex linear mapping $Df(x_0) : E \to F$ such that*

$$\lim_{h \to 0} \frac{\|f(x_0 + h) - f(x_0) - Df(x_0)(h)\|_F}{\|h\|_E} = 0.$$

The mapping $Df(x_0)$ is the derivative *of f at x_0. A function $f : U \to F$ is* holomorphic in U *if it has derivative at each point of U. A holomorphic function is* biholomorphic *if it has a holomorphic inverse. Such a function is called a* biholomorphic mapping.

In the case $E = \mathbf{C}^m$ and $F = \mathbf{C}^n$ this notation agrees with the usual definition of holomorphic functions.

Let F^* denote the *dual* of the Banach space F. The set F^* consists of all continuous complex linear mappings $F \to \mathbf{C}$. The norm

$$\|x^*\|_{F^*} = \sup\{|x^*(x)| \mid \|x\|_F \leq 1\}$$

makes F^* a Banach space.

Definition 2.4.7 *A set $A \subset F^*$ is* total *if the implication*

$$\forall y^* \in A : y^*(x) = 0 \Rightarrow x = 0$$

holds.

The following criterium (cf. [14]) is useful when having to check whether a mapping between Banach spaces is holomorphic.

Theorem 2.4.7 *A function $f : U \to F$ is holomorphic if and only if it satisfies one of the following conditions:*

- *For every $x \in U$ and $e \in E$, the function $z \mapsto f(x + ze)$ is a holomorphic function on an open neighborhood of the origin with values in F.*

- *The function $f : U \to F$ is continuous and there exists a total subset $A \subset F^*$ such that*

$$\forall y^* \in A : \ y^* \circ f : U \to \mathbf{C} \text{ is holomorphic.}$$

Using this characterization of holomorphic functions and an explicit singular integral expression for quasiconformal mappings one can prove the following result which has applications in Chapter 5. A clear proof for the following can be easily derived from the proof of Theorem 5.1 in [60, page 207].

Theorem 2.4.8 *Let U be the upper half-plane and μ a Beltrami differential. Let f^{μ} be the unique quasiconformal mapping whose complex dilatation agrees a. e. with μ and whose extension to the closure of U keeps* 0, 1 *and* ∞ *fixed. Then, for every $z \in U$, the mapping $\mu \mapsto f^{\mu}(z)$ is a holomorphic complex valued function on the unit ball of L^{∞}.*

Chapter 3

Geometry of Riemann surfaces

3.1 Introduction to Chapter 3

It is the aim of this book to show how the elementary considerations of Chapter 1 can be applied to Riemann surfaces and Teichmüller spaces. Many profound results concerning the famous *moduli problem* can be reduced to the results of Chapter 1. This is surprising, because considerations of Chapter 1 were mostly based on multiplication of matrices and studying traces of matrices and words of matrices.

In order to be able to see how this is done we have to review some results of the topology of surfaces, Riemann surfaces and the uniformization of Riemann surfaces. Much of the material of this Chapter will be presented without proofs. We will, however, give references to proofs at all places where complete reasoning is omitted.

Our main references to topology of surfaces are the monographs of Edwin Spanier ([91]) and Lars Ahlfors and Leo Sario ([3]). Even though we assume that the basic results are known we will here review everything that is necessary for our considerations.

3.2 Riemann and Klein surfaces

A connected topological Hausdorff space Σ is *a surface with boundary* if the following condition is satisfied: every point $p \in \Sigma$ has an open neighborhood U which is homeomorphic to an open set in the closed upper half plane $\{z|\operatorname{Im} z \geq 0\}$. A homeomorphism $z : U \to z(U)$ is called *a local variable* at the point $p \in U$. The pair (U, z) is called *a coordinate chart.*

Points $p \in \Sigma$ for which all the local charts (U, z), $p \in U$, satisfy $z(p) \in \mathbf{R}$

are called *boundary points* of Σ. Recall that, by the definition, $z(U) \subset \{z| \operatorname{Im} z \geq 0\}$ for all local charts (U, z). Boundary points of Σ form *the boundary* $\partial\Sigma$ of Σ. By *a surface* we usually mean a surface with an empty boundary. Surfaces that are allowed to have boundary are referred to as surfaces with boundary.

In the sequel we have to deal with homeomorphisms of open sets of the closed upper half–plane. Let $f : A \to f(A)$ be such an homeomorphism. We say that it is *holomorphic* or *analytic* if it is holomorphic in the usual sense in $A \cap \{z| \operatorname{Im} z > 0\}$.

A homeomorphism $f : A \to f(A)$ is called *dianalytic* if either f itself or the complex conjugate of f is holomorphic in each component of the set A.

Let $z : U \to z(U)$ and $w : V \to w(V)$ be two local variables such that $U \cap V \neq \emptyset$. Then we may form the mapping $z \circ w^{-1} : w(V \cap U) \to z(V \cap U)$. This mapping is called *the coordinate transition function.*

Definition 3.2.1 *A collection* $\mathcal{U} = \{(U_i, z_i)|i \in I\}$ *of coordinate charts of* Σ *is* an atlas *of the surface* Σ *if* $\Sigma = \cup_{i \in I} U_i$. *An atlas* $\mathcal{U}$ *is:*

orientable *if each coordinate transition function* $z_i \circ z_j^{-1}$ *is an orientation preserving homeomorphism.*

dianalytic *if all the coordinate transition functions are dianalytic homeomorphisms.*

complex analytic *if all the coordinate transition functions are holomorphic homeomorphisms.*

Two complex analytic atlases $\mathcal{U}$ and $\mathcal{V}$ are called *equivalent* if $\mathcal{U} \cup \mathcal{V}$ is a complex atlas as well. An equivalence relation is introduced in the same manner for the other types of atlases.

An equivalence class of complex atlases is called *a complex structure* of the surface Σ. Dianalytic, and orientable structures are defined in the same way as equivalence classes of the respective structures.

A surface Σ which has an orientable structure is called *orientable.* An orientable structure of an orientable surface determines an *orientation.*

Definition 3.2.2 A Riemann surface *is a topological surface* Σ *together with a complex structure* X. A Klein surface *is a topological surface, possibly with boundary,* Σ *together with a dianalytic structure* Y.

Observe that Riemann surfaces, as defined above, do not have boundary points and that each Riemann surface automatically is a Klein surface as well. A Klein surface (Σ, X) that cannot be made into a Riemann surface is called *non–classical* while orientable Klein surfaces with empty boundary are called *classical.*

A classical Klein surface always carries two Riemann surface structures which are complex conjugates, or mirror images, of each other.

It is well known that every surface with boundary can be made into a Klein surface, i.e., every such surface carries dianalytic structures. It is also well known that every orientable surface can be made into a Riemann surface. These results follow from the topological fact that every surface with or without boundary can be represented as a branched covering of the Riemann sphere or the unit disk (for a proof in the case of compact surfaces see e.g. [8, Theorem 1.7.2, page 49]). We will not prove this result here because we do not use it anywhere. The problem that we are concerned about is to parametrize the set of all analytic or dianalytic structures of a given surface in some reasonable way. This is also the famous moduli problem.

3.3 Elementary surfaces

Simpliest compact topological surface is the sphere S^2 in $\mathbf{R}^3$

$$S^2 = \{(x, y, z) \mid x^2 + y^2 + z^2 = 1\}.$$

Identifying the antipodal points (x, y, z) and $(-x, -y, -z)$ on the sphere S^2 one obtains the *real projective plane* $\mathbf{P}^1(\mathbf{R})$. This surface is not anymore orientable in the sense that it has only one side. It is not possible to embedd the real projective plane in $\mathbf{R}^3$.

The *torus* T is the quotient surface

$$T = \mathbf{C}/\langle z \mapsto z + 1,\, z \mapsto z + i\rangle.$$

Here $\langle z \mapsto z+1,\, z \mapsto z+i\rangle$ is the group generated by the elements $z \mapsto z+1$ and $z \mapsto z + i$.

Let R be the strip

$$R = \{z \mid 0 \leq \text{ Im } z \leq 1\}.$$

The torus can be obtained from a rectangle identifying its opposite sides. If we identify only one pair of opposite sides we get an annulus

$$A = R/\langle z \mapsto z + 1\rangle.$$

The identification of a pair of opposite sides can also be done changing the orientation. That is achieved by the mapping $z \mapsto \text{Re } z + 1 + (1 - \text{Im z})i$. The quotient surface

$$M = R/\langle z \mapsto \text{Re } z + 1 + (1 - \text{Im z})i\rangle$$

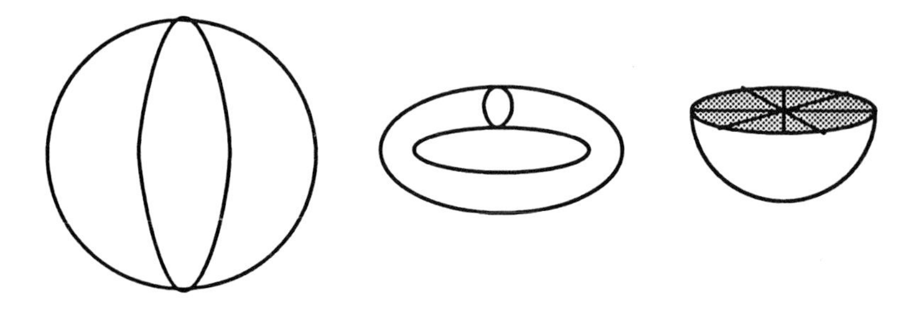

Figure 3.1: The sphere, the torus and the real projective plane. The figure shows the real projective plane cut open along an arc in such a way that the cutting yields a half–sphere. The straight lines indicate how antipodal boundary points have to be identified in order to get a 'real' real projective plane.

is the well known *Möbius strip.* It is the only one–sided, i.e., a non–orientable, surface that can be embedded in an euclidean 3–space. Figure 3.2 shows the Möbius strip in the upper left hand corner.

The *Klein bottle* is another famous non–orientable surface. The usual way to picture the Klein bottle is to consider first an ordinary bottle from which a small open disk is deleted from the bottom. This is actually an annulus. To get the Klein bottle, identify the two boundary components of the bottle (i.e. the annulus) we started with in such a way that the orientation gets reversed in the process. In this way one gets a one–sided bottle.

Another way to picture the Klein bottle is shown in figures 3.2, 3.3 and 3.4. All these illustrations are due to Ari Lehtonen. In these figures the Klein bottle is formed by taking two copies of the Möbius strip and identifying the boundary points. Figure 3.2 illustrates this. The resulting surface is some kind of a twisted product of a figure–8 curve and the unit circle. Figure 3.3 illustrates this.

3.4 Topological classification of surfaces

Section 3.3 gives a rather concrete picture about some elementary non–orientable surfaces. We can make this a little bit more precise by considering curves on surfaces. *A curve* α on Σ is the image of the closed unit interval $I = [0, 1]$ under a continuous mapping $\alpha : I \to \Sigma$. We often use the same notation α for a mapping $I \to \Sigma$ and for its image in Σ. The points $\alpha(0)$ and $\alpha(1)$ are *the end–points* of α. Observe that the orientation of the unit

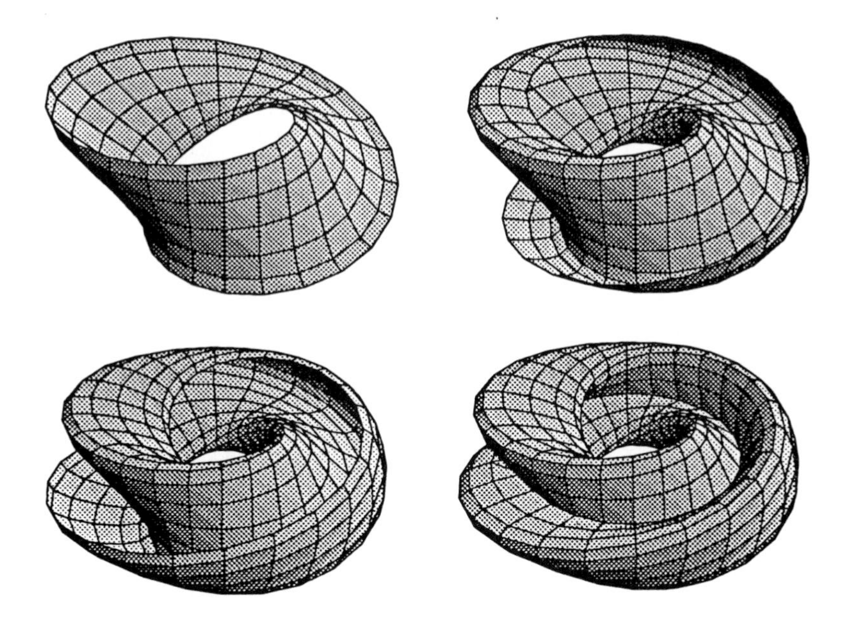

Figure 3.2: Simpliest non–orientable surface with boundary components is the famous *Möbius strip*. It can be embedded in a 3–space. Gluing two Möbius strips together along the boundary curves yields the *Klein bottle*, which cannot be embedded into a 3–space. This illustration shows first the Möbius strip which is then gradually enlarged to a projection of the Klein bottle in a 3–space. The projection necessarily cuts itself. We thank Ari Lehtonen for this picture.

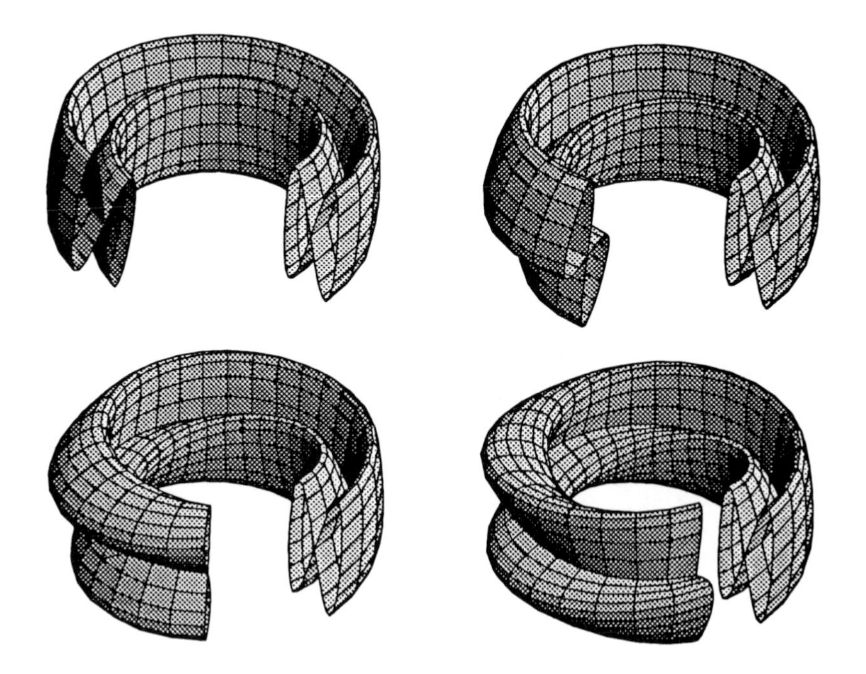

Figure 3.3: The Klein bottle can be viewed as a twisted product of a figure–8 curve and the unit circle.

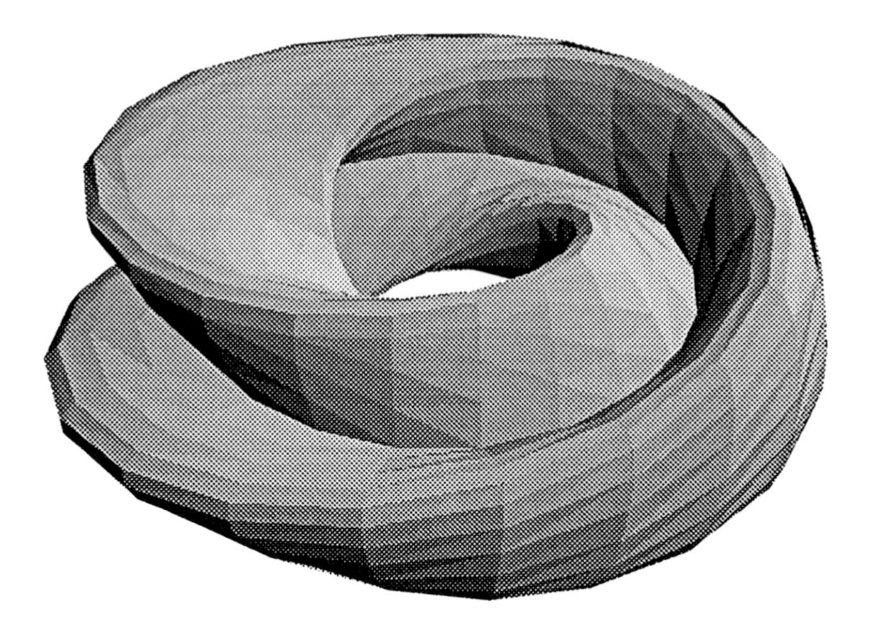

Figure 3.4: The Klein Bottle.

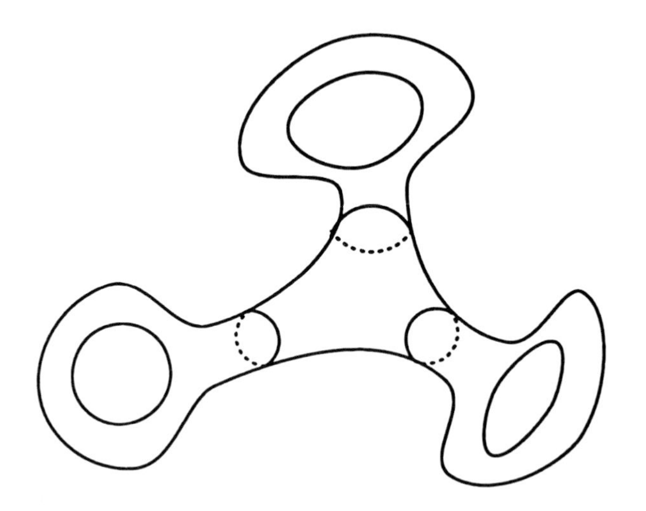

Figure 3.5: A surface of genus g is a sphere with g handles.

interval $[0,1]$ induces an orientation for each curve $\alpha : [0,1] \to \Sigma$.

A curve α is *closed* if its end–points agree. A closed curve α is called *simple* if the mapping α is injective when restricted to the half–open interval $[0,1[$. A simple closed curve α is *two–sided* if there exists an open and connected set U such that $\alpha \subset U$ and $U \setminus \alpha$ has two components. A curve α is *one–sided* if it is not two sided.

Another characterization for orientable surfaces can be obtained by means of simple closed curves: A surface Σ is orientable if and only if every simple closed curve in (the interior of) Σ is two–sided. We will not prove this result here and we do not need it in our applications.

The *Euler characteristic* $\chi(\Sigma)$ of a compact surface Σ is defined (for more details see e.g. [91, p. 113 and p. 172] or [3]) in terms of finite triangulations of Σ. If n_j is the number of j–faces of a finite triangulation of Σ, $j = 0,1,2$, then

$$\chi(\Sigma) = n_0 - n_1 + n_2.$$

For an orientable compact surfaces Σ, the Euler characteristic is an even number and the number $g = g(\Sigma) = 1 - \chi(\Sigma)/2$ is called *the genus* of the orientable surface Σ.

The genus g of an orientable surface Σ is always non–negative. A compact surface of genus 0 is homeomorphic to the Riemann sphere $\hat{\mathbf{C}}$.

A compact orientable surface of genus g, $g > 0$, can be thought of as a sphere with g handles.

Let now Σ be a surface, orientable or not. Let $p \in \Sigma$. We proceed and define next *the orientable covering* Σ^o *of* the surface Σ as follows.

An atlas $\mathcal{U} = \{(U_i, z_i) | i \in I\}$ of Σ is *maximal* if the following condition is satisfied:

Let $\mathcal{V}$ be an atlas of Σ such that $\mathcal{U} \cup \mathcal{V}$ is also an atlas of Σ. Then $\mathcal{U} \subset \mathcal{V}$.

Assume now that $\mathcal{U} = \{(U_i, z_i) | i \in I\}$ is a *maximal* atlas of Σ.

Form first the disjoint union

$$S = \cup_{i \in I} U_i.$$

Let (U_i, z_i) and (U_j, z_j) be two charts such that $U_i \cap U_j \neq \emptyset$. Let $p \in U_i \cap U_j$. Being a disjoint union the set S has two points which both correspond to the point $p \in \Sigma$, namely the point $p \in U_i$ and the same point $p \in U_j$. To make a distinction between these points, call the latter one p'. Next we identify the points p and p' if the corresponding coordinate transition function $z_i \circ z_j^{-1}$ is orientable. This gives us the set Σ^o.

We still have to define a topology for the set Σ^o. That is done via the natural projection $\pi : \Sigma^o \to \Sigma$. A topology for Σ^o is defined requiring $\pi : \Sigma^o \to \Sigma$ be locally homeomorphic. Then Σ^o is clearly an orientable surface. It is connected if and only if Σ is non–orientable. For an orientable Σ, Σ^o has two components which are both homeomorphic to Σ.

If $\partial\Sigma \neq \emptyset$, then Σ^o has two points lying over each boundary point of Σ. Identifying these two points gives us a surface Σ^c which is called *the complex double* of the surface Σ. The covering $\pi : \Sigma^o \to \Sigma$ induces a mapping $\pi : \Sigma^c \to \Sigma$ which is a ramified double covering of Σ. It is a local homeomorphism at all points $p \in \Sigma^c$ for which $\pi(p) \notin \partial\Sigma$. At points lying over the boundary of Σ, the projection is a folding similar to the mapping $x + iy \mapsto x + i|y|$ at the real axis. For more details about this mapping see [8, 1.6].

In this way we form the complex double Σ^c of a surface with boundary Σ. If Σ is orientable and $\partial\Sigma = \emptyset$, then Σ^c has two components which are both homeomorphic to Σ. For all other surfaces Σ, Σ^c is a connected orientable surface without boundary.

Observe that the covering group of the branched covering $\pi : \Sigma^c \to \Sigma$ is generated by an orientation reversing involution $\sigma : \Sigma^c \to \Sigma^c$. (An *involution* is a mapping whose square is the identity.)

Also $\Sigma = \Sigma^c/\langle\sigma\rangle$, where $\langle\sigma\rangle$ is the group generated by σ. Here $\Sigma^c/\langle\sigma\rangle$ is *the quotient surface* obtained by identifying p with $\sigma(p)$ for all $p \in \Sigma^c$.

Above we defined the genus of a classical compact surface. *The genus*, or, more precisely, *the arithmetic genus* of a non–classical compact surface Σ is defined as the genus of the complex double Σ^c of Σ.

When speaking of the genus of non–classical surfaces we always mean this arithmetic genus. Observe that in some other text books the genus of a non–classical surface has another meaning.

Since the Euler characteristic of a circle vanishes, we have $\chi(\Sigma^c) = 2\chi(\Sigma)$ for all non–classical compact surfaces Σ. Therefore we conclude that for non–classical compact surfaces $\chi(\Sigma) = 1 - g$. Recall that for classical surfaces Σ, $\chi(\Sigma) = 2 - 2g$.

Compact topological surfaces are classified topologically by the following parameters:

- the genus $g = g(\Sigma)$ of Σ,
- the number $n = n(\Sigma)$ of components of $\partial\Sigma$,
- *the index of orientability* $k = k(\Sigma)$ which is defined setting $k = 0$ for orientable surfaces Σ and $k = 1$ for non–orientable surfaces Σ.

It is clear that the above parameters g, n and k are topological invariants in the sense that for surfaces, that are homeomorphic to each other, the genus, the number of boundary components and the index of orientability agree.

These parameters are also related to each other as described in the following theorem [96].

Theorem 3.4.1 *The index of orientability k and the number n of boundary components of a compact genus g surface Σ with boundary components satisfy:*

- *If $k = 0$ then $n \equiv g + 1 \pmod 2$ and $n > 0$.*
- *If $k = 1$ then $0 \leq n \leq g$.*

These are the only restrictions and all possible configurations of g, n and k satisfying these conditions appear as invariants of some compact genus g surface with boundary components.

Corollary 3.4.2 *There are*

$$\left\lfloor \frac{3g+4}{2} \right\rfloor \tag{3.1}$$

topologically different non–classical compact genus g surfaces with boundary components.

These are classical results of Weichhold. They can also be found in the works of Klein ([45], [47], [49]). We will not prove this result here. In Section 3.7 we will, however, construct a partial proof for Theorem 3.4.1. In that section we show a way to construct all the surfaces of Theorem 3.4.1. What remains not shown in this monograph is that there are no other surfaces. That can be found in the works referred above.

Observe that classifying non–classical topological compact genus g surfaces is the same thing as classifying all orientation reversing involutions σ of a classical compact genus g surface. This follows from the above constructions. The complex double Σ^c of a non–classical genus g surface Σ is a branched covering of Σ and the cover group is generated by a single orientation reversing involution $\sigma : \Sigma^c \to \Sigma^c$. Then $\Sigma = \Sigma^c/\langle\sigma\rangle$.

Conversely, given a classical compact surface Σ' of genus g and an orientation reversing involution $\sigma : \Sigma' \to \Sigma'$, then $\Sigma'/\langle\sigma\rangle$ is a compact non–classical surface of genus g.

Definition 3.4.1 A symmetric surface (Σ, σ) *is a classical surface* Σ *together with an orientation reversing involution* $\sigma : \Sigma \to \Sigma$. *The involution* σ *is* the symmetry *of the symmetric surface* Σ.

For a symmetry $\sigma : \Sigma \to \Sigma$ define $n(\sigma)$ as the number of the components of the fixed–point set Σ_σ of σ. Define also $k(\sigma)$ setting

$$k(\sigma) = 2 - \text{number of components of } \Sigma \setminus \Sigma_\sigma.$$

It is immediate that the invariants $n(\sigma)$ and $k(\sigma)$ are simply the corresponding invariants n and k of the quotient surface $\Sigma/\langle\sigma\rangle$.

We say that two symmetries σ and τ of a topological surface Σ are *conjugate to each other* if there exists a homeomorphism $f : \Sigma \to \Sigma$ such that $\sigma \circ f = f \circ \tau$. This is equivalent to the condition that the surfaces $\Sigma/\langle\sigma\rangle$ and $\Sigma/\langle\tau\rangle$ are homeomorphic to each other.

Corollary 3.4.2 implies then that the number of different conjugacy classes of symmetries of a classical genus g surface is given by formula (3.1).

For a later application it is useful to be able to have a concrete understanding of the topology of a compact surface. To that end we will now describe an explicit way of building a compact surface starting with certain elementary surfaces.

We use the three elementary surfaces, the sphere S^2, the torus T and the real projective plane $\mathbf{P}^1(\mathbf{R})$, to construct concrete models of more complicated surfaces.

Let p and k be non–negative integers, and let S^2_{p+k} denote the complement of the union of $n+k$ open disks on S^2 whose closures are disjoint. Let T_1 denote the complement of an open disk on T. Likewise $\mathbf{P}^1(\mathbf{R})_1$ is the complement of an open disk on $\mathbf{P}^1(\mathbf{R})$.

The Euler characteristics are:

- $\chi(S^2_{p+k}) = 2 - p - k$.
- $\chi(\mathbf{P}^1(\mathbf{R})_1) = 0$.
- $\chi(T_1) = -1$.

Let A and B be two compact subsurfaces with boundary of a topological surface. The Euler characteristics satisfy

$$\chi(A \cup B) = \chi(A) + \chi(B) - \chi(A \cap B). \tag{3.2}$$

Build a surface Σ^p_k by taking S^2_{p+k} and gluing to it p copies of the surface T_1 to it along the boundary components and k copies of the surface $\mathbf{P}^1(\mathbf{R})_1$ also along the boundary components. In this way one obtains a compact surface Σ^p_k without boundary components.

Using (3.2) and the fact that the Euler characteristic of a circle vanishes we compute that

$$\chi(\Sigma^p_k) = 2 - p - k + p\chi(T_1) = 2 - 2p - k.$$

In this way one can build all compact surfaces without boundary components starting from the sphere, the torus and the real projective plane. We say that the surface Σ^p_k is the *direct sum* of the sphere, p torii and k real projective planes. These real projective planes are also called *cross caps* of the surface Σ^p_k. Any compact surface with boundary components can then be obtained from a surface of type Σ^p_k by deleting a suitable number of open disks with disjoint closures.

Let $\Sigma^{p,n}_k$ be a compact surface with p handles (i.e. torii), k cross caps and n boundary components. Using the above computation we easily get the formula

$$\chi(\Sigma^{p,n}_k) = 2 - 2p - k - n$$

for the Euler characteristic of the surface $\Sigma^{p,n}_k$.

3.5 Discrete groups of Möbius transformations

A topological group G is a group G together with a topology for which the inverse $G \to G$, $g \mapsto g^{-1}$ and the group operation $G \times G \to G$ are continuous mappings.

Let G now be a group of Möbius–transformations. Elements of G are of the form

$$g_k(z) = \frac{a_k z + b_k}{c_k z + d_k}, \; ad - bc = +1, \; a, b, c, d \in \mathbf{C}. \tag{3.3}$$

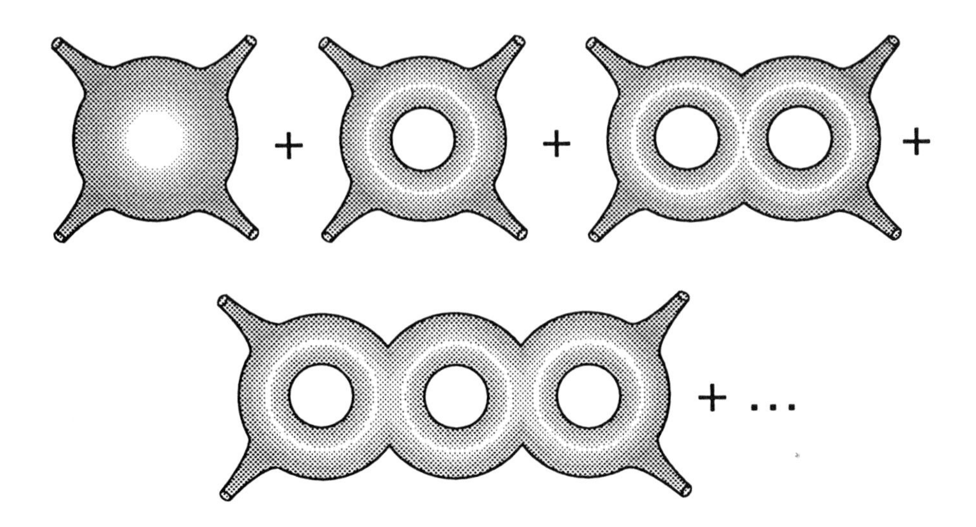

Figure 3.6: Any compact topological surface can be built combining a finite number of simple structures. This figure shows a few first steps of such a construction. The holes shown in the figure may be boundary components of the resulting surface or may be used to add cross caps or may be used to add more structure. We thank Ari Lehtonen for this illustration.

We can define a topology on G by saying that a sequence $(g_k,\ k = 1, 2, \ldots)$ converges to

$$g_\infty(z) = \frac{a_\infty z + b_\infty}{c_\infty z + d_\infty}$$

if and only if we *may* choose representations of the type (3.3) for the elements g_k in such a way that $a_k \to a_\infty$, $b_k \to b_\infty$, $c_k \to c_\infty$ and $d_k \to d_\infty$ as $k \to \infty$. This is the usual definition. Together with this topology G is a topological group.

Recall that a topological space is called *discrete* if all of its subsets are open. Likewise we say that a topological group G is *discrete* if all of its subsets are open. Observe especially that for a discrete topological group all subsets consisting of one point only are open. In the applications of the methods presented in Chapter 1, discrete groups of Möbius transformations play an important role.

A related concept to discreteness is *discontinuity*. We say that a group G whose elements are Möbius transformations *acts discontinuously* at a point $x \in \hat{\mathbf{C}}$ if the following holds:

- The *stabilizer* of G at x,

  $$G_x = \{g \in G \mid g(x) = x\}$$

  is finite.

- There is an open set $U \subset \hat{\mathbf{C}}$, $x \in U$, such that $g(U) = U$ for all $g \in G_x$ and $g(U) \cap U = \emptyset$ for all other elements of G.

The set of points where G acts discontinuously is called the *set of discontinuity,* or *regular set*, and is usually denoted by $\Omega = \Omega(G)$. It follows from the definition that Ω is an open G–invariant subset of the extended complex plane.

A group of Möbius transformations that acts discontinuously in some domain D is sometimes said to be *properly discontinuous.*

Definition 3.5.1 *A group G of Möbius transformations is a* Kleinian group *if $\Omega(G) \neq \emptyset$.*

Lemma 3.5.1 *A Kleinian group is either finite or countable.*

Proof. Choose a point $z \in \Omega(G)$ such that the stabilizer of z, G_z, is trivial. Then

$$G(z) = \{g(z) \mid g \in G\}$$

is a discrete set. Therefore $G(z)$ is either finite or countable.

Observe finally, that the cardinality of the set $G(z)$ is the same as the cardinality of the group G.

Definition 3.5.2 *A Kleinian group G is a* Fuchsian group *if the following holds: there is a disk (or a half-plane) D of the extended complex plane $\hat{\mathbf{C}}$ such that each element of the group G maps D onto itself and $D \subset \Omega(G)$.*

The following result is immediate by definitions:

Lemma 3.5.2 *Every Kleinian group is discrete.*

The converse is not true. The *Picard group*

$$P = \{z \mapsto \frac{az+b}{cz+d} \mid ad - bc = 1 \text{ and } a, b, c, d \in \mathbf{Z}[i]\}$$

is clearly discrete but not discontinuous because it can be shown that for any $z \in \hat{\mathbf{C}}$, the set $\{g(z) \mid g \in P\}$ is dense in $\hat{\mathbf{C}}$.

Theorem 3.5.3 *Let G be a group of Möbius transformations mapping the unit disk D onto itself. Then G is discontinuous if and only if it is discrete.*

Proof. As we have already observed, a discontinuous group is clearly also discrete. It suffices, therefore, to show that a discrete group of Möbius transformations mapping the unit disk onto itself is also discontinuous.

To prove this assume that the group G is not discontinuous at some point $z_0 \in D$. This means that we can find infinite sequences $z_1, z_2, \ldots \in D$ and $g_1, g_2, \ldots \in G$ such that $z_n \to z_0$ as $n \to \infty$ and $g_n(z_n) = z_0$ for each n.

Set now

$$A_n(z) = \frac{z - z_n}{1 - \overline{z_n} z}, \qquad z \in D, \ n = 0, 1, \ldots$$

and

$$C_n = A_{n+1} \circ g_{n+1}^{-1} \circ g_n \circ A_n^{-1}, \qquad n = 1, 2, \ldots.$$

Since $C_n(0) = 0$, we conclude by Schwarz's lemma that

$$C_n(z) = \lambda_n z, \qquad |\lambda_n| = 1.$$

Thus, by passing to a subsequence if necessary, we may assume that $\lambda_n \to \lambda_0$ as $n \to \infty$. This means that the sequence C_n converges to C_0 as $n \to \infty$.

The points z_n are assumed to be distinct. Therefore also the elements g_n of the group G are distinct. The mapping $h_n = g_{n+1}^{-1} \circ g_n$ maps z_n onto z_{n+1} for each n. Assume that infinitely many of the mappings h_n agree. Then we may as well suppose that $h_n = h$ for all n. We conclude that

$$z_0 = \lim_{n \to \infty} z_n = \lim_{n \to \infty} h(z_n) \tag{3.4}$$

is a fixed point of the mapping h, which is, therefore, an elliptic element of the group G. But since h is elliptic, (3.4) can happen if and only if $z_n = z_0$

for large enough values of n. But this is not possible, since we assumed that all the points z_n are distinct. This implies that the mappings $h_n = g_{n+1}^{-1} \circ g_n$ are distinct elements of the group G.

On the other hand, the above considerations imply that

$$h_n = A_{n+1} \circ C_n \circ A_n^{-1} \to A_0^{-1} \circ C_0 \circ A_0 \text{ as } n \to \infty.$$

This is not possible since the group G is discrete.

Observe that in the above proof the contradiction was derived from the assumption that there is one point $z_0 \in D$ at which the group G does not act discontinuously.

Above proof actually implies the following stronger statement:

Theorem 3.5.4 *Assume that the Möbius group G leaves the unit disk invariant. The following conditions are equivalent:*

1. $D \subset \Omega(G)$.
2. $D \cap \Omega(G) \neq \emptyset$.
3. *G is discrete.*

In this monograph we are concerned with Möbius groups that act either in the unit disk or in the upper half–plane. By Theorem 3.5.3 discontinuity and discreteness are equivalent properties for such groups.

Let G be a discrete Möbius group that acts in the upper half–plane U. Assume that the stabilizer of $z \in U$ in G is trivial. Form the set

$$D_z(G) = \{z' \in U \mid d(z', z) < d(g(z'), z) \, \forall g \in G, \, g \neq 1\}.$$

It is clearly an open set and it has the following properties:

1. No two points of $D_z(G)$ are equivalent under the action of the group G.
2. For every point $w \in U$ there exists an $g_w \in G$ such that $g_w(w) \in D_z(G)$.
3. The relative boundary of $D_z(G)$ in U consists of piecewise analytic arcs.
4. For every arc $a \subset \partial D_z(G)$ there is an arc $a' \subset \partial D_z(G)$ and an element $g \in G$ such that $g(a) = a'$.

We will not show here that $D_z(G)$ satisfies these properties. A detailed proof can be found in the monograph of Alan F. Beardon [10, §9.4., pp. 226 - 234].

Definition 3.5.3 *By a* fundamental domain *a Fuchsian group G acting in the upper half-plane U we mean an open subset $D(G)$ of U satisfying the above conditions 1 – 4. The above defined fundamental domain $D_z(G)$, for a point $z \in U$ that is not a fixed-point of a non-identity element of G, is called the* Dirichlet *or the* Poincaré polygon *for G.*

3.6 Uniformization

For any surface Σ we may form *the universal covering surface* $\tilde{\Sigma}$ which is simply connected and admits *a projection* $\pi : \tilde{\Sigma} \to \Sigma$ that is a local homeomorphism. Furthermore each point $p \in \Sigma$ has a neighborhood U_p such the restriction of the projection π to each component of $\pi^{-1}(U_p)$ is a homeomorphism between the component and U_p.

Homeomorphic self–mappings g of $\tilde{\Sigma}$ satisfying $\pi \circ g = \pi$ form *the cover group* G of the universal covering $\pi : \tilde{\Sigma} \to \Sigma$.

Since π is a local homeomorphism, the group G acts *discontinuously* on $\tilde{\Sigma}$. The action is also *free* in the sense that no non–identity element of G has fixed–points in $\tilde{\Sigma}$.

The cover group G of the universal covering of Σ has the property that if $\pi(p) = \pi(q)$, for $p, q \in \tilde{\Sigma}$, then there exists an $g \in G$ such that $g(p) = q$. It follows that $\Sigma = \tilde{\Sigma}/G$. This is, of course, quite standard.

More details concerning the universal cover can be found, for instance, in [3].

Let X be a dianalytic structure of the surface Σ. Then, by requiring the mapping π be locally analytic, we may lift the dianalytic structure of Σ to a dianalytic structure $\tilde{X}$ to $\tilde{\Sigma}$.

Next observe that, since $\tilde{X}$ is simply connected, it is orientable. Therefore the dianalytic structure $\tilde{X}$ of $\tilde{\Sigma}$ is induced by a some complex structure Y. Hence we may suppose that $\tilde{X}$ is a *complex structure* such that $\pi : (\tilde{\Sigma}, \tilde{X}) \to (\Sigma, X)$ is dianalytic. There are actually two possible complex structures $\tilde{X}$ satisfying this condition. They are complex conjugates of each other.

We conclude that any Klein surface (Σ, X) has a Riemann surface as its universal covering surface. From the equation $\pi \circ g = \pi$ and from the fact that $\pi : (\tilde{\Sigma}, \tilde{X}) \to (\Sigma, X)$ is dianalytic and a local homeomorphism, it follows that each element g of the cover group G is a dianalytic self mapping of $(\tilde{\Sigma}, \tilde{X})$.

The following result is the famous Riemann mapping theorem:

Theorem 3.6.1 *A simply connected Riemann surface without boundary is either the extended complex plane, the finite complex plane or the upper half plane U.*

We will not prove this result here. A proof can be found, for instance, in the monograph of Farkas and Kra [29, Theorem IV.4.4, p. 182].

By this theorem we may suppose that, for any Klein surface (Σ, X), *the interior* of the universal covering $(\tilde{\Sigma}, \tilde{X})$ is one of the standard Riemann surfaces of Theorem 3.6.1. Then, by the preceding observation, we conclude that elements of a cover group G corresponding to a Klein surface (Σ, X) are all Möbius transformations. If the Klein surface (Σ, X) is not orientable, then the group G necessarily contains orientation reversing Möbius transformations.

Analyzing all possible groups that act properly discontinuously on the Riemann sphere or on the finite complex plane we conclude that:

- $(\tilde{\Sigma}, \tilde{X})$ is the Riemann sphere if and only if (Σ, X) is either the Riemann sphere itself or the real projective plane. In this case the Euler characteristic of Σ is positive.

- $(\tilde{\Sigma}, \tilde{X})$ is the finite complex plane if and only if (Σ, X) is one of the following surfaces:

  - finite complex plane,
  - torus,
  - infinite cylinder,
  - Klein bottle.

  In this case the Euler characteristic of Σ vanishes.

If the Euler characteristic of Σ is negative, then the universal covering $(\tilde{\Sigma}, \tilde{X})$ of (Σ, X) is always the upper half–plane together with certain intervals on the real axis. These intervals correspond to the boundary of the Klein surface (Σ, X). Then the upper half–plane itself is the universal covering of the interior of the Klein surface (Σ, X). It is sometimes technically easier to consider the interior of a surface instead of the whole surface with boundary. By abuse of language, we may later speak of the upper half–plane as the universal covering of a Klein surface which may have boundary components.

Let $Q \in \Sigma$ be a point and α, β closed curves on Σ with end–points at Q. We say that α and β are *homotopic* if there exists a continuous mapping $h : [0,1] \times [0,1] \to \Sigma$ such that $h(0,s) = h(1,t) = Q$ and $h(s,0) = \alpha(s)$,

$h(t,1) = \beta(t)$, for all $s,t \in [0,1]$. We use the notation $\alpha \approx \beta$ to indicate that α and β are homotopic to each other.

It is obvious that $\approx$ is a equivalence relation in the set of closed curves with end–points at Q. The corresponding set of equivalence classes, or *homotopy classes*, of closed curves at Q is denoted by $\pi_1(\Sigma, Q)$.

If α and β are closed curves as above, then their product $\alpha\beta$ is defined setting

$$\alpha\beta(t) = \begin{cases} \beta(2t) & \text{for } 0 \le t \le \frac{1}{2} \\ \alpha(2t-1) & \text{for } \frac{1}{2} < t \le 1 \end{cases}$$

It is rather straightforward to verify that this multiplication determines a multiplication in $\pi_1(\Sigma, Q)$ and that $\pi_1(\Sigma, Q)$ is then a group. It is called *the fundamental group* or *the first homotopy group* of the surface Σ at the *base point* Q.

The definition of the fundamental group $\pi_1(\Sigma, Q)$ depends on the choice of the base–point Q. The choice of this base–point is, however, irrelevant. Standard arguments show that if Q' is another point of Σ then the groups $\pi_1(\Sigma, Q)$ and $\pi_1(\Sigma, Q')$ are isomorphic to each other. This isomorphism can be constructed in the following way. Let first γ be a curve such that $\gamma(0) = Q$ and $\gamma(1) = Q'$. Then define

$$\pi_1(\Sigma, Q) \to \pi_1(\Sigma, Q'),\ [\alpha] \mapsto [\gamma\alpha\gamma^{-1}]. \tag{3.5}$$

This is a well defined mapping and an isomorphism. The isomorphism (3.5) depends, of course, on the choice of the connecting curve γ. An other choice of γ changes the isomorphism (3.5) by an inner automorphism of $\pi_1(\Sigma, Q')$.

Let G be the covering group of the universal cover of the surface Σ. There is an almost canonical morphism

$$i : \pi_1(\Sigma, Q) \to G, \tag{3.6}$$

which is defined in the following way. Choose first a point $\tilde{Q} \in U$ (or $\in \mathbf{C}$ or $\in \hat{\mathbf{C}}$) lying over the point $Q \in \Sigma$. Every closed curve γ with end points at Q can be lifted to the universal covering space U (or $\mathbf{C}$ or $\hat{\mathbf{C}}$) of (Σ, X). The lifting becomes unique when we require that its starting point is the previously fixed point $\tilde{Q}$. Let $\tilde{Q}_1$ be the end-point of this lifting. Then also $\tilde{Q}_1$ is a point over Q and hence there is an element g_γ of G such that $g_\gamma(\tilde{Q}) = \tilde{Q}_1$. The element g_γ defined by this condition is unique because non–identity elements the group G do not have fixed–points in D.

Lemma 3.6.2 *The Möbius transformation $g_\gamma \in G$ depends only on the homotopy class of γ.*

Proof. This is a standard result in topology and holds even in a more general setting. The result follows from the 'homotopy lifting property' of the universal cover and the discontinuity of the action of the cover group G. For a proof we refer to [91, Corollary 8 on page 88].

By Lemma 3.6.2 $[\gamma] \to g_\gamma$ is a well–defined mapping $\pi_1(\Sigma) \to G$. The following result gives us the inverse of this morphism and thus proves that it is actually an isomorphism between $\pi_1(\Sigma, Q)$ and G.

Lemma 3.6.3 *Assume that (Σ, X) is a Klein surface such that the universal cover of the interior of (Σ, X) is the unit disk D with the cover group G. Let α be a closed curve representing a point of $\pi_1(\Sigma, Q)$. Let $g_\alpha = i([\alpha])$ be the element fo the group G which corresponds to $[\alpha] \in \pi_1(\Sigma, Q)$. Then each curve in D with end–points z and $g_\alpha(z)$, $z \in D$, projects to a closed curve on Σ that is homotopic to the curve α.*

Proof. This is also quite standard and follows from rather general topological arguments. To prove the result we have to construct a homotopy between the projected curve and the original curve α. That can be done using hyperbolic geometry.

Choose now a point $z \in D$ and a lifting of the curve α to a curve $\tilde{\alpha}$ in D. Then $\tilde{\alpha}$ is a continuous mapping $I \to D$ satisfying $\alpha = \pi \circ \tilde{\alpha}$, where $\pi : D \to \Sigma$ is the projection. Let $\beta : I \to D$ be a curve in D such that $\beta(0) = z$ and $\beta(1) = g_\alpha(z)$.

We form first a continuous mapping $F : I \times I \to D$ in the following way. Let $(t, s) \in I \times I$. Define $F(t, s)$ as the point on the hyperbolic geodesic between $\beta(t)$ and $\tilde{\alpha}(t)$ which divides that geodesic in the ratio $s : (1 - s)$. Then clearly $F(t, 0) = \beta(t)$, $F(t, 1) = \tilde{\alpha}(t)$ for all $t \in I$. We have, furthermore $g_\alpha(F(0, s) = F(1, s)$ for all $s \in I$. This implies that $\pi \circ F : I \times I \to \Sigma$ is a homotopy between the closed curves α and $\pi \circ \beta$ proving the lemma.

By the above result we conclude that $\pi_1(\Sigma, Q) \to G$, $[\alpha] \mapsto g_\alpha$, is an isomorphism. It depends on the choice of the point $\tilde{Q} \in D$ lying over the base–point Q. Another choice of $\tilde{Q}$ changes the corresponding isomorphism by an inner automorphism of G.

Definition 3.6.1 *We say that the transformation $g_\alpha \in G$ of Lemma 3.6.3* covers *the (homotopy class of the) curve α.*

Next we recall a result (cf. e.g. [91, statement 12 on page 149]) describing the fundamental group of a surface. Assume that Σ is compact surface with n boundary components, p handles and k cross–caps.

Theorem 3.6.4 *The fundamental group $\pi_1(\Sigma, Q)$ of the surface Σ is generated by the elements $\alpha_1, \beta_1, \ldots, \alpha_p, \beta_p$ (which correspond to the handles of Σ), $\gamma_1, \ldots, \gamma_n$ (which correspond to the boundary components) and $\delta_1, \ldots, \delta_k$ (which correspond to cross-caps) satisfying the relation*

$$\prod_{j=1}^{p}[\alpha_j, \beta_j] \prod_{i=1}^{k} \delta_k^2 \prod_{l=1}^{n} \gamma_l = 1. \tag{3.7}$$

Recall that, in the above theorem, $[\alpha_j, \beta_j]$ is the commutator of α_j and β_j.

Now let X be a dianalytic structure on Σ and assume that (the interior of) the Klein surface (Σ, X) is D/G for a reflection group G.

The group G is isomorphic to $\pi_1(\Sigma, Q)$. Let $i : \pi_1(X, Q) \to G$ be an isomorphism and let $g_j = i(\alpha_j)$, $h_j = i(\beta_j)$, $d_j = i(\gamma_j)$ and $s_j = i(\delta_j)$ be the elements of G corresponding to the generators of $\pi_1(X, Q)$. Then the set

$$\mathcal{K} = \{g_1, h_1, \ldots, g_p, h_p, d_1, \ldots, d_n, s_1, \ldots, s_k\}$$

generates G and satisfies

$$\prod_{j=1}^{p}[g_j, h_j] \prod_{i=1}^{k} s_j^2 \prod_{l=1}^{n} d_j = 1. \tag{3.8}$$

Definition 3.6.2 *Generators g_j, h_j, d_i and s_l of the group G satisfying the relation (3.8) are called the* standard generators *for G.*

Observe that if X is a compact classical Riemann surface of genus p, then there are no generators of type d_j or s_l. The standard generators for such a group are Möbius transformations g_1, $h_1, \ldots, g_p$, h_p satisfying the single relation

$$\prod_{j=1}^{p}[g_j, h_j] = 1. \tag{3.9}$$

This is also the most complicated case since, from the relation (3.9), it is not possible to solve any one of the generators in terms of the other generators. But if $n > 0$, then the group G is actually freely generated, since, in this case, one can solve one of the elements d_j by the relation (3.8). If $n = 0$ but $k > 0$, then we have minor technicalities to take care of. In this case we can express, by the relation (3.8), one of the Möbius transformations s_j^2 in terms of the other generators. By the construction, s_j's are now orientation reversing Möbius transformations mapping the unit disk onto itself. Such a Möbius transformation is a glide reflection and its square is a hyperbolic Möbius transformation.

Recall then (cf. considerations on page 16) that the hyperbolic Möbius transformation s_j^2 alone determines the glide reflection s_j uniquely. Therefore, even if the group G in this case is not freely generated, we know everything about the generator s_1 when all the other generators are given. The classical case of compact and oriented Riemann surfaces remains the most complicated one.

Let $X = (\Sigma, X)$ and $Y = (\Sigma, Y)$ be Klein surfaces. We need to consider continuous mappings between X and Y. Assume that $X = D/G$ and $Y = D/G'$, where G and G' are Fuchsian groups. Let $f : X \to Y$ be a continuous mapping. It can then be lifted to a continuous mapping $F : D \to D$ satisfying

$$F \circ \pi' = f \circ \pi \tag{3.10}$$

where $\pi : D \to X$ and $\pi' : D \to Y$ are the projections. If F is one such lifting, then also $g' \circ F \circ g$, $g \in G$, $g' \in G'$, is a lifting of f, i.e., a continuous mapping satisfying (3.10). All continuous liftings of a continuous mapping f are obtained from one lifting in this manner.

The equation (3.10) implies to the following observation:

For each $z \in D$ and for each $g \in G$ there exists an $g'_z \in G'$ such that

$$F(g(z)) = g'_z(F(z)). \tag{3.11}$$

Lemma 3.6.5 *The Möbius transformation g'_z in equation (3.11) does not depend on the point $z \in D$.*

Proof. Let z_0 and z_1 be two points of D and let $\alpha : I \to D$ be a curve such that $\alpha(0) = z_0$ and $\alpha(1) = z_1$. Denote $\alpha(t) = z_t$. Then we may define the element g'_{z_t} for each $t \in I$ by the equation (3.11). Recall that since the group G' acts freely in D the element g'_{z_t} is uniquely defined by (3.11).

It suffices to show that $g'_{z_0} = g'_{z_1}$. To that end, let $I' = \{t \in I \mid g'_{z_t} = g'_{z_0}$. Clearly $I' \neq \emptyset$, since $0 \in I'$. By the discontinuity of the group G' we then conclude that both I' and its complement are open in I. This implies then that $I' = I$ proving the lemma.

The above result follows also from the discreteness of the group G. To see this, consider the above defined elements $g_{z_t} \in G$. They give a continuous mapping $I \to G$, $t \mapsto g_{z_t}$. But since G is discrete and I connected, such a continuous mapping is necessarily constant. The difficulty in this reasoning is to show the continuity of the mapping $t \mapsto g_{z_t}$.

By Lemma 3.6.5 a continuous lifting $F : D \to D$ of a continuous mapping $f : D/G \to D/G'$ defines a morphism $f^{\#} : G \to G'$, setting $f^{\#}(g) = g'_z$ for any $z \in D$. This morphism does not depend on the choice of the point $z \in D$.

Definition 3.6.3 *The homomorphism $f^{\#} : G \to G'$ defined above is said to be* induced *by the mapping f.*

Observe that one continuous mapping induces many homomorphisms between the corresponding group. An induced morphism depends, of course, on the choice of the lifting F. Another choice F' induces a homomorphism $(f')^{\#}$ which is obtained from $f^{\#}$ by composing it with an inner automorphism of G'.

Definition 3.6.4 *Homomorphisms $i : G \to G'$ and $j : G \to G'$ are called* equivalent *if there exists an element $g_0' \in G'$ such that $i(g) = g_0' \circ j(g) \circ (g_0')^{-1}$, holds for all $g \in G$, i.e., if i is obtained composing j with an inner automorphism of G'.*

We conclude now that all continuous liftings of a continuous mapping $f : D/G \to D/G'$ induce equivalent homomorphisms $G \to G'$.

In our applications we are mainly interested in homeomorphisms $f : D/G \to D/G'$. Assume now that f is a homeomorphism. The above construction that was done for the mapping f can just as well be applied to the mapping f^{-1}. If $F : D \to D$ is a lifting of $f : D/G \to D/G'$, then F^{-1} is a lifting of f^{-1}. Using these liftings in the constructions for $f^{\#}$ and $(f^{-1})^{\#}$ we conclude easily that

$$(f^{-1})^{\#} = (f^{\#})^{-1}.$$

This means that for a homeomorphism $f : D/G \to D/G'$ the induced homomorphism $f^{\#}$ is an isomorphism.

Lemma 3.6.6 *Assume that $f : D/G \to D/G'$ and $g : D/G \to D/G'$ are homotopic homeomorphisms. Then they induce equivalent isomorphisms $f^{\#} : G \to G'$ and $g^{\#} : G \to G'$.*

Proof. Let $H : (D/G) \times I \to D/G'$ be a continuous mapping such that $H(p,0) = f(p)$ and $H(p,1) = g(p)$ for all $p \in D/G$. This homotopy between the mappings f and g can be lifted to a continuous mapping

$$\tilde{H} : D \times I \to D$$

such that $\tilde{H}(\cdot,0)$ is a lifting of the mapping f and $\tilde{H}(\cdot,1)$ is that of g. The mapping $\tilde{H}$ is, furthermore, compatible with the action of the groups G and G' in the sense that for each $t \in I$ and for each $g \in G$ there exists an $g_t' \in G'$ for which $\tilde{H}(g(z),t) = g_t'(H(z,t))$ for each $z \in D$.

It suffices to show that this element g_t' does not depend on t. Then it follows that $\tilde{H}(\cdot,0)$ and $\tilde{H}(\cdot,1)$ induce the same isomorphism. Repeating the reasoning of the proof of Lemma 3.6.5 we see that this follows from the discontinuity of the action of the group G'.

Lemma 3.6.7 *Assume that $f : D/G \to D/G'$ and $g : D/G \to D/G'$ are two homeomorphisms such that the induced isomorphisms $f^{\#}$ and $g^{\#}$ are equivalent. Then the mappings f and g are homotopic to each other.*

Proof. Assume that the isomorphism $f^{\#}$ is defined by the lifting F of the mapping f and $g^{\#}$ by the lifting G of g. The assumption that the isomorphisms $f^{\#}$ and $g^{\#}$ are equivalent means that there exists a Möbius transformation $g_0' \in G'$ such that

$$f^{\#}(g) = g_0' \circ g^{\#}(g) \circ (g_0')^{-1} \tag{3.12}$$

for all $g \in G$.

Since G is a lifting of g, then such is $G' = g_0' \circ G$ as well. Let us see which isomorphism is induced by this lifting. To that end write

$$\begin{aligned} G'(g(z)) &= g_0' \circ G(g(z)) \\ &= g_0' \circ g^{\#}(g) \circ G(z) \\ &= g_0' \circ g^{\#}(g) \circ (g_0')^{-1} \circ g_0' \circ G(z) \\ &= g_0' \circ g^{\#}(g) \circ (g_0')^{-1} \circ G'(z) \end{aligned}$$

to conclude, by (3.12), that the lifting $G' = g_0' \circ G$ of g induces the *same* isomorphism than the lifting F of the mapping f. We may, therefore, assume that F and G are liftings of f and g, respectively, inducing the same isomorphism $f^{\#} : G \to G'$.

Next we construct a homotopy $\tilde{H} : D \times I \to D$ between the mappings F and G defining $\tilde{H}(z,t)$ as that point on the geodesic arc from $F(z)$ to $G(z)$ which divides this arc in ratio $t : (1-t)$. Then $\tilde{H}$ is a continuous mapping and $\tilde{H}(\cdot, 0) = F(\cdot)$, $\tilde{H}(\cdot, 1) = G(\cdot)$.

Using the fact that F and G define the same isomorphism $f^{\#} : G \to G'$ and the fact that this isomorphism is an isometry of the hyperbolic metric, we then conclude that

$$\tilde{H}(g(z), t) = f^{\#}(g) \circ \tilde{H}(z, t) \tag{3.13}$$

for all $z \in D$ and for all $t \in I$. Equation (3.13) implies that $\tilde{H}$ induces a homotopy between the mappings $f : D/G \to D/G'$ and $g : D/G \to D/G'$.

Lemma 3.6.7 has the following corollary that will be applied later.

Lemma 3.6.8 *Assume that G is a Fuchsian group acting in the upper half–plane U and that hyperbolic non–identity elements of G have at least three fixed–points on $\mathbf{R} \cup \{\infty\}$. Then if $f : U/G \to U/G$ and $g : U/G \to U/G$ are homotopic to each other and holomorphic, then $f = g$.*

Proof. Under the assumptions of the lemma, $h = f \circ g^{-1}$ is holomorphic and homotopic to the identity. Let $H : U \to U$ be a lifting of h. Since H is a holomorphic homeomorphism, it is a Möbius transformation and can be immediately extended to the whole infinite complex plane. Let us do that.

Observe that the identity mapping $U \to U$ is, of course, a lifting of the identity mapping $U/G \to U/G$. By Lemma 3.6.7 H and the identity mapping induce equivalent isomorphisms $G \to G$. Repeating a part of argument of the proof of Lemma 3.6.7 we may suppose that H induces the *same* isomorphism $G \to G$ as the identity mapping, i.e., that H induces the identity $G \to G$.

This means that for all $g \in G$, $z \in U$ and $n \in \mathbf{N}$ we have

$$H(g^n(z)) = g^n(H(z)). \tag{3.14}$$

Assume that g is hyperbolic and fix a point $z \in U$. Consider the equation (3.14) for various values of n. Recall that

$$\lim_{n \to +\infty} g^n(z) = a(g) \text{ and } \lim_{n \to -\infty} g^n(z) = r(g)$$

where $a(g)$ is the attracting fixed–point of g and $r(g)$ is the repelling fixed–point.

From equation (3.14) it then follows that $H(a(g)) = a(g)$ and $H(r(g)) = r(g)$. This applies to all hyperbolic elements of G. Since hyperbolic elements of G have at least three different fixed–points. We conclude that the orientation preserving Möbius transformation H fixes three points. It is, therefore, the identity mapping.

3.7 Models for symmetric surfaces

Even though we did not prove Theorem 3.4.1 we will, in this section, give a concrete construction that shows us the existence of all the symmetric surfaces of Theorem 3.4.1.

Consider first involutions σ with index of orientability $k(\sigma) = 0$. Let n be an integer with $g - (n-1) = g + 1 - n$ even. Take a Riemann surface of genus $(g+1-n)/2$. Delete n open disks from it. Assume that the disks are chosen in such a manner that their closures are disjoint. Then one gets a Riemann surface Y of genus $(g+1-n)/2$ with n boundary components.

Let $\overline{Y}$ denote the Riemann surface obtained from Y by replacing the complex structure of Y with its conjugate structure, i.e. by replacing all local variables z with their complex conjugates $\overline{z}$. $\overline{Y}$ is simply the mirror image of Y. Glue the Riemann surfaces Y and $\overline{Y}$ together identifying the boundary points. In that way one gets a compact Riemann surface X of genus g. The identity mapping $Y \to \overline{Y}$ induces an antiholomorphic involution $\sigma : X \to X$ such that the curves of X corresponding to the boundary curves of Y remain point-wise fixed. Therefore the parameters of Theorem 3.4.1 satisfy $n(\sigma) = n$ and $k(\sigma) = 0$ for this involution. This is how one can construct topologically all symmetries σ of a genus g Riemann surface X satisfying $k(\sigma) = 0$ and $n(\sigma) \equiv g + 1 (\mathrm{mod}\ 2)$.

Let α be a closed curve left point–fixed under the above involution σ. Let A be a tubular neighborhood of α. Then the universal covering of A is the strip

$$\tilde{A} = \{z \in \mathbf{C} | -1 < \mathrm{Im}\, z < 1\}.$$

Furthermore we may suppose that σ maps A onto itself and that the complex conjugation is a lifting of $\sigma : A \to A$ onto $\tilde{A}$. Then the real axis covers the curve α.

Everything here is only topological. So assuming that the covering group of $\tilde{A} \to A$ is generated by $z \mapsto z + 2$ we do not restrict the generality.

Define the function $H : \tilde{A} \to \tilde{A}$ setting $H(x + iy) = (x + 1 - y + iy)$. Then the complex conjugation $\tau(x + iy) = x - iy$ and $H \circ \tau$ are both self–mappings of $\tilde{A}$. Both of them map the real axis onto itself but only the complex conjugation keeps it point–wise fixed.

Let $f_\alpha : X \to X$ be defined setting $f_\alpha(p) = p$ for $p \in X \setminus A$. In A define f_α as the mapping induced by $H : \tilde{A} \to \tilde{A}$. The mapping $f_\alpha : X \to X$ defined in this way is clearly a homeomorphism.

Definition 3.7.1 *The mapping $f_\alpha : X \to X$ is the* Dehn twist *of X along the curve α.*

It is easy to check that $f_\alpha \circ \sigma$ is also an involution of X. For this involution we have $k(f_\alpha \circ \sigma) = 1$ and $n(f_\alpha \circ \sigma) = n(\sigma) - 1$. Figure 3.7 illustrates how the involution $f_\alpha \circ \sigma$ maps a curve that intersects the curve α.

Repeating this procedure for each component of the fixed–point set of σ we can clearly in this way construct topologically any symmetric surface for which $n < n(\sigma)$ and $k = 1$. This how one can construct topological models for all symmetric Riemann surfaces.

Dehn twist was here used to construct topological models of symmetric surfaces. It is an important concept and has many applications. The rather formal definition given above can be replaced by the following shorter definition which relies partly on reader's geometric intuition.

To give this definition for the Dehn twist, orient first the simple closed curve α in some way. You have two choices, they both lead to the same deformation.

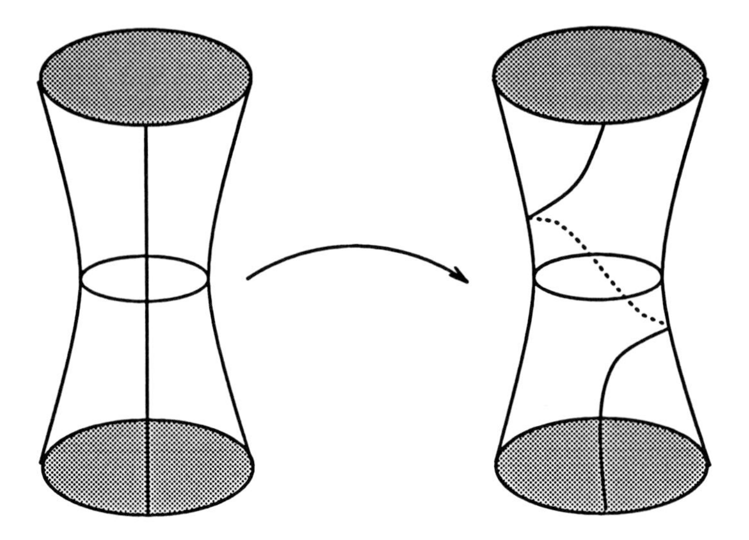

Figure 3.7: The twisted involution $f_\alpha \circ \sigma$ does not keep the curve α pointwise fixed.

The orientation of the curve α tells us the positive direction along α, and, since X is assumed to be oriented, it also separates the left hand side of α from the right hand side. Now cut the surface X open along α. You obtain in this way a new surface X' which has two boundary components α' and α'' corresponding to the curve α. These boundary components can be thought of as corresponding to the left hand side of α, call that one α', and to right hand side of α. Turn the left hand side of α, i.e., the curve α' full turn around in the positive direction of α and glue it back to the curve α''. In this way we obtain the topology of the surface X is not changed. But curves crossing α get replaced by curves crossing α and going once around 'the handle' corresponding to α. This is also called *the left Dehn twist.* It is easy to check that the definition of the left Dehn twist does not depend on the orientation of the curve α.

3.8 Hyperbolic metric of Riemann surfaces

By the Uniformization Theorem, every Riemann surface W (or, more generally, every Klein surface) with negative Euler characteristic can be expressed as $W = D/G$, where G is a discontinuous group consisting of Möbius transformations mapping the hyperbolic unit disk onto itself.

We equip the the unit disk D by the hyperbolic metric of constant curvature -1. For the definition and basic properties of this metric see Appendix A. By Theorem A.2.1 in Appendix A, the elements of G are isometries of the hyperbolic metric of D. We then conclude that the hyperbolic metric of the unit disk D projects to a metric of the surface W. We call this metric *the hyperbolic metric* of W. The hyperbolic metric of a Riemann surface is a complete metric of constant curvature -1 (cf. (A.4)).

Let W now be a compact hyperbolic Riemann (or Klein) surface with boundary curves $\alpha_1, \ldots, \alpha_p$. On W we often use *the intrinsic hyperbolic metric* in which the boundary components are geodesic curves of a finite length (see e.g. [1, page 45]). This metric can be obtained in the following way. First form the Schottky double of W by gluing W and its mirror image $\overline{W}$ together along the boundary components. *The intrinsic metric* of W is the restriction of the usual hyperbolic metric of the Schottky double of W to W itself.

Let W be a compact Riemann or Klein surface. The well–known Gauß–Bonnet formula (cf. [34, Page 228]) gives the following expression for the hyperbolic area $A(W)$ of the surface W

$$A(W) = -2\pi\chi(W). \tag{3.15}$$

We will not prove this result here. It will, nevertheless, play an important role in our considerations.

The area formula (3.15) implies that the area of a compact Riemann or Klein surface depends only on the topological type of the surface and *not* on the complex or dianalytic structure.

A Riemann surface is usually thought of as being a sphere with handles. The hyperbolic metric then tells us how thick and how long the handles are. If the surface degenerates in such a way that some handles become very long, then they necessarily become thin at the same time. This is a loose observation can be understood by the invariance of the hyperbolic area. We will later give a completely different proof for this fact using the considerations of Chapter 1.

From now on we assume that every compact Riemann or Klein surface without boundary is equipped with the hyperbolic metric. Surfaces with boundary components are — unless otherwise stated — assumed to be equipped with the *intrinsic hyperbolic metric.*

A main goal of this chapter, and also of this monograph, is to provide tools that allow us to make precise the above loose remarks concerning the degeneration of Riemann surfaces.

3.9 Hurwitz Theorem

For a later application we discuss here a classical construction which leads to an estimate concerning the order of the automorphism group of a compact classical Riemann surface. This result is known as the *Hurwitz Theorem* (see [42]). Assume that X and Y are classical Riemann surfaces and that $f : X \to Y$ is a non-constant holomorphic function.

For any point $p \in X$ we may choose local variables z at p and w at $f(p)$ such that $z(p) = w(f(p)) = 0$. Then $w \circ f \circ z^{-1}$ is a holomorphic function defined in a neighborhood of the origin where it has series expansion of the form

$$w \circ f \circ z^{-1}(\zeta) = \sum_{k=n}^{\infty} a_k \zeta^k,$$

where n is a positive integer and $a_n \neq 0$.

A closer analysis then reveals that we may actually choose the local variables z and w in such a way that

$$w \circ f \circ z^{-1}(\zeta) = \zeta^n. \tag{3.16}$$

Definition 3.9.1 *If $n > 1$ in (3.16), then $p = z^{-1}(0)$ is a* branch point *of f. The number n is referred to as the* ramification number *of f at p and f is said to have* multiplicity *n at p. The number $b_f = b_f(p) := n - 1$ is called the* branch number of *f at p.*

Lemma 3.9.1 *Let $f : X \to Y$ be a non-constant holomorphic mapping between classical compact Riemann surfaces. Let $q \in Y$. The number*

$$m_f := \sum_{p \in f^{-1}(q)} (b_f(p) + 1)$$

is finite and does not depend on the choice of the point q.

Proof. We reproduce here an argument of [29, page 12]. For each integer $n \geq 1$, define

$$Y_n = \{q \in Y \mid \sum_{p \in f^{-1}(q)} (b_f(p) + 1) \geq n\}.$$

Using the form (3.16) for f we conclude that Y_n is open for each n.

Let us show that each Y_n is also closed. Therefore fix a positive integer n and assume that the points $q_1, q_2, \ldots$ belong to Y_n and that the series (q_k) converges to a point $q \in Y$. We have to show that also $q \in Y_n$.

If $q_k = q$ for some value of k, then q trivially belongs to Y_n and we have nothing to prove. Assume that $q_k \neq q$ for all k. Then we may also assume that the points q_k are distinct.

A non–constant holomorphic function between compact Riemann surfaces can have at most finitely many branch points. Therefore, by passing to a subsequence, we may also suppose that none of the points q_k is a branch point of f. This means that, for each k, $f^{-1}(q_k)$ consists of at least n distinct points $p_1^k, p_2^k, \ldots, p_n^k$ of X. Since X is compact we may suppose, again by passing to a subsequence, that each sequence p_j^k converges to some p_j^∞ as $k \to \infty$. Of course it may happen that the points p_j^∞ are not anymore distinct.

Now clearly $f(p_j^\infty) = q$ for each j. It also follows that

$$\sum_{p \in f^{-1}(q)} b_f(p) + 1 \geq n$$

even if the points p_j^∞ are not all distinct. The set Y_n is therefore closed.

Since Y is connected, each Y_n is either empty or all of Y. For large enough values of n, Y_n is clearly empty. We conclude therefore, that

$$m_f = \sum_{p \in f^{-1}(q)} (b_f(p) + 1) = \sup\{n \mid Y_n \neq \emptyset\}$$

is independent of the point q.

Definition 3.9.2 *Let $f : X \to Y$ be a non–constant holomorphic function between compact Riemann surfaces. The* total branching number *B_f of f is defined setting*

$$B_f = \sum_{p \in X} b_f(p).$$

This is finite, since $b_f(p)$ vanishes for all but finitely many points $\in X$.

Theorem 3.9.2 (Riemann–Hurwitz Relation) *Let $f : X \to Y$ be a non–constant holomorphic function between compact Riemann surfaces. Assume that the genus of X is g and that of Y is γ. Then*

$$g = n(\gamma - 1) + 1 + B_f/2, \tag{3.17}$$

where n is the degree of f (the cardinality of $f^{-1}(q)$ is n for almost all $q \in Y$).

For a detailed proof we refer to [42, V.2] or to [29, page 19], which both use the same argument. Equation (3.17) follows by considering a suitable triangulation of Y that lifts to a triangulation of X. Then one can express the Euler characteristics of X and Y in a way that immediately yields (3.17).

Theorem 3.9.3 *Let X be a compact Riemann surface of genus g, $g > 1$. Then the group* $\mathrm{Aut}(X)$ *of holomorphic automorphisms of X is finite and has at most* $84(g-1)$ *elements.*

Proof. This is classical and due to Adolf Hurwitz ([43, II.7]). First observe that obviously $\mathrm{Aut}(X)$ acts properly discontinuously on X. Therefore, $X/\mathrm{Aut}(X)$ is a compact Riemann surface and the projection $\pi : X \to X/\mathrm{Aut}(X)$ is a non–constant holomorphic mapping of a compact Riemann surface of genus > 1 onto a compact Riemann surface. Such a mapping has a finite degree N. Clearly this number N is also the number of elements of $\mathrm{Aut}(X)$, i.e., the order of $\mathrm{Aut}(X)$. We conclude that $\mathrm{Aut}(X)$ has finite order N.

Now $\pi : X \to X/\mathrm{Aut}X$ is a non–constant holomorphic mapping whose branch points $p_1, p_2, \ldots, p_r$ are also fixed–points of non–identity elements of $\mathrm{Aut}(X)$. Let $\mathrm{Aut}(X)_{p_j}$ be the stabilizer of p_j in $\mathrm{Aut}(X)$. Denote by ν_j the number of elements of $\mathrm{Aut}(X)_{p_j}$. The branching order of π at p_j is clearly

$$b_\pi(p_j) = \nu_j - 1 = \mathrm{ord}\mathrm{Aut}(X)_{p_j} - 1.$$

Next observe that there are N/ν_j distinct points of X which are equivalent to p_j under the action of $\mathrm{Aut}(X)$. Therefore the total branch number of π is

$$B = \sum_{j=1}^{r} \frac{N}{\nu_j}(\nu_j - 1) = N \sum_{j=1}^{r}(1 - \frac{1}{\nu_j}).$$

The Riemann–Hurwitz relation (3.17) yields now

$$2g - 2 = N(2\gamma - 2) + N \sum_{j=1}^{r}(1 - \frac{1}{\nu_j}), \tag{3.18}$$

where γ is the genus of $X/\mathrm{Aut}(X)$.

The proof follows now from a detailed analysis of all possibilities of (3.18).

Case I: $\gamma \geq 2$. In this case (3.18) implies

$$2g - 2 \geq 2N \text{ or } N \leq g - 1.$$

Case II: $\gamma = 1$. In this case (3.18) yields

$$2g - 2 = N \sum_{j=1}^{r}(1 - \frac{1}{\nu_j}). \tag{3.19}$$

If $r = 0$, then $g = 1$ which is contrary to our assumptions. Therefore $r \geq 1$. Equation (3.19) then implies that

$$2g - 2 \geq \frac{1}{2}N \text{ or } N \leq 4(g-1).$$

Case III: $\gamma = 0$. This is the only case where we may have $N = 84(g-1)$. Repeating the above arguments several times finally proofs the theorem. We skip the detailed computation of this case. It can be found in [29, pp. 243 – 244].

By Theorem 3.9.3 the order of the automorphism group of a compact Riemann surface of genus g is $84(g-1)$. It is an interesting and difficult problem to find out whether this limit can be achieved for a given genus g.

By explicit computations Klein showed that the automorphism group of a genus 2 Riemann surface has at most 48 elements. Then Gordan ([35]) showed that the order of the automorphism group of a genus 4 Riemann surface is, at most, 120. Wiman showed in [100] and in [99] in the cases of genus 5 and 6 Riemann surfaces these limits are 192 and 420, respectively. Therefore the bound $84(g-1)$ is not attained in the cases of genus 2, 4, 5 or 6 Riemann surfaces.

In the case of genus 3 Riemann surfaces this bound is attained: Klein's quartic

$$x^3y + y^3 + x = 0$$

defines a Riemann surface of genus 3 with the maximal number $168 = 84(3-1) = 84(g-1)$ automorphisms. This is the only Riemann surface of genus 3 with this property.

This detailed analysis of Riemann surfaces of a low genus indicates that the maximal order $84(g-1)$ for the group of automorphisms is rarely achieved. That is the case, indeed. One should observe, however, that it is achieved by infinitely many genera g. Recently Ravi Kulkarni has investigated the (large finite) groups that are automorphism groups of some compact Riemann surfaces. For more details we refer to [55] and to references given there.

This theory concerning the order and the structure of automorphism groups of compact Riemann surfaces can also be generalized to compact non–classical Riemann surfaces. For a detailed account of these extensions of the classical theory see the monograph of Emilio Bujalance, José J. Etayo, José M. Gamboa and Grzegorz Gromadzki [17]

3.10 Horocycles

The geometry of hyperbolic Klein surfaces X can be studied by using the corresponding group G for which $X = D/G$. Closed geodesic curves are closely related with the transformations of the group G. To understand this relation we start with considering groups G that contain parabolic Möbius transformations.

Definition 3.10.1 *A Möbius transformation g is said to be a* primitive *element of a Fuchsian group G, if the following holds:*

$$g = h^n, \ h \in G \Leftrightarrow n = \pm 1.$$

Lemma 3.10.1 *Assume that G is a Fuchsian group acting in the upper half-plane U. Let g be a parabolic element and h any non-identity element of G such that the fixed-point of g is also a fixed-point of h. Then there are integers m and n such that*

$$g^m = h^n. \tag{3.20}$$

Proof. Assume that h is hyperbolic. By conjugation we may assume that g is either $g(z) = z + b$ and $h(z) = kz$ for some $b \neq 0$, k, $k > 0, \neq 1$. The set of the elements $g_n(z) = h^{-n} \circ g \circ h^n(z) = z + bk^{-n}$, $n \in \mathbf{Z}$, contains then a sequence of distinct elements of G converging to the identity. This is not possible. Therefore h can not be hyperbolic.

Assume next that h is parabolic. By conjugation we may assume again that both parabolic elements g and h have ∞ as fixed point. Then they are of the form $g(z) = z + b$ and $h(z) = z + b'$. The statement of the Lemma 3.10.1 (equation (3.20)) is equivalent to saying that b/b' is rational. But if this ratio were irrational, then

$$\{mb + nb' \mid m, n \in \mathbf{Z}\}$$

would be dense in $\mathbf{R}$ and one could easily build a sequence of distinct elements of G converging to the identity. That is not possible since G is assumed to be discrete.

Lemma 3.10.2 *Suppose that G is a Fuchsian group acting in the upper half-plane U and containing the parabolic transformation $g(z) = z + 1$ as a primitive element. Let $H = \{z \mid \operatorname{Im} z > 1\}$ and assume that $h \in G$. Then either $h(H) \cap H = \emptyset$ or $h = g^n$ for some $n \in \mathbf{Z}$. H is the largest half-plane having this property.*

Proof. Assume that

$$h(z) = \frac{az+b}{cz+d}, \; ad-bc=1, \text{ and } h \neq g^n \text{ for any } n \in \mathbf{Z}.$$

If $c = 0$, then h has ∞ as fixed point. By our present assumptions and by Lemma 3.10.1 this is not possible. Therefore we conclude that $|c| > 0$.

We have to show that $h(H) \cap H = \emptyset$. Let $h_1 = h \circ g \circ h^{-1}$ and define inductively

$$h_{k+1} = h_k \circ g \circ h_k^{-1}.$$

Write

$$h_k(z) = \frac{a_k z + b_k}{c_k z + d_k}, \; a_k d_k - b_k c_k = 1.$$

Then

$$\begin{aligned} a_{k+1} &= 1 - a_k c_k \\ b_{k+1} &= a_k^2 \\ c_{k+1} &= -c_k^2 \\ d_{k+1} &= 1 + a_k c_k. \end{aligned}$$

If $|c| < 1$, then $c_k = -c^{2^k} \to 0$. This implies that $a_k \to 1$, $b_k \to 1$ and $d_k \to 1$ as $k \to \infty$. Therefore we conclude that in this case $h_k \to g$. Since $|c| > 0$, elements h_k are also distinct. This is not possible since G was assumed to be discrete. Hence $|c| \geq 1$.

Let $z \in H$. Then $\operatorname{Im} h(z) = (\operatorname{Im} z)/|cz+d|^2$. Provided that h is not a power of g, $\partial H' = h(\partial H)$ is a circle tangent to $\mathbf{R}$ at a/c. We compute:

$$\begin{aligned} \text{diameter of } H' &= \sup_{x \in \mathbf{R}} \left| \frac{a(x+i)+b}{c(x+i)+d} - \frac{a}{c} \right| \\ &= \sup_{x \in \mathbf{R}} \frac{1}{|c|^2} \left| \frac{1}{x+i+(d/c)} \right| \\ &= \frac{1}{|c|^2} \\ &\leq 1 \end{aligned}$$

which implies the first statement of the lemma.

The *elliptic modular group* $\mathrm{SL}_2(\mathbf{Z})$ provides an example which shows that H is the largest half–plane for which the first statement of the lemma holds. For the basic properties of the elliptic modular group we refer to the discussion in Chapter 5. A fundamental domain for this group is described on page 182. This remark completes the proof of the lemma.

Definition 3.10.2 *Using the notation of Lemma 3.10.2, we say that H is a* horocycle *at the fixed point of g.*

We can improve the above result if we assume that the group G does not contain elliptic elements. Here we follow an argument of [81] and [90]. Assume that this is the case, i.e., that G is a Fuchsian group which acts freely in the upper half-plane U.

Let $g \in G$, $g(\infty) \neq \infty$, and let $I(g)$ denote the isometric circle of g. For the definition of the isometric circle see Section 1.4. The center of $I(g)$ lies on the real axis, $g(I(g)) = I(g^{-1})$, and $I(g)$ and $I(g^{-1})$ have the same radius. If g is parabolic, then $I(g)$ and $I(g^{-1})$ are tangent to each other at the the fixed point of g, otherwise $I(g) \cap I(g^{-1}) = \emptyset$.

Lemma 3.10.3 *The difference $g(z) - z$ is real for a point z in the upper half-plane if and only if $z \in I(g)$.*

This lemma follows directly from the geometry of the action of the Möbius transformation g and the definition of the isometric circle. Proof is left to the reader.

Lemma 3.10.4 *Let $g \in G$ be such that $g(\infty) \neq \infty$. If G contains the translation $g_\omega : z \mapsto z + \omega$, then $(g(z) - z)/\omega$ is not an integer for any $z \in U$.*

Proof. Suppose that $g(z) - z = n\omega$ for some integer n and $z \in U$. Then $g^{-1} \circ g_\omega^n$ fixes z. It follows that $g^{-1} \circ g_\omega^n$ is elliptic, which is not possible by our assumptions.

Lemma 3.10.5 *Suppose that $g_\omega : z \mapsto z + \omega$ is in G. If $g \in G$ does not fix ∞, then the radius r_g of $I(g)$ satisfies $r_g \leq \omega/4$.*

Proof. By geometry

$$\max_{z \in I(g)} |g(z) - z| - \min_{z \in I(g)} |g(z) - z| = 4r_g. \tag{3.21}$$

Now $(g(z) - z)/\omega$ is real on $I(g)$ by Lemma 3.10.3. If the total variation of $(g(z)-z)/\omega$ along $I(g)$ were more than 1, then $(g(z)-z)/\omega$ would necessarily take an integer value at some point in $I(g)$. By Lemma 3.10.4 this is not possible. We conclude, therefore, that

$$\max_{z \in I(g)} \frac{|g(z) - z|}{\omega} - \min_{z \in I(g)} \frac{|g(z) - z|}{\omega} \leq 1. \tag{3.22}$$

Inequality (3.22) together with equation (3.21) implies now the lemma.

Theorem 3.10.6 *Assume that the group G contains, besides the identity, only hyperbolic and parabolic Möbius transformations mapping the upper half-plane onto itself. Assume further that $g_1(z) = z + \omega$ is a primitive element of the group G. Let $g \in G$ be such that $g(\infty) \neq \infty$. If $\operatorname{Im} z > \omega/4$, then $\operatorname{Im} g(z) < \omega/4$. The number $\omega/4$ is the smallest possible.*

Proof. The first part of the statement follows directly from Lemma 3.10.5. To prove that $\omega/4$ is the smallest number with this property, assume that $\omega = 1$ and let g_0 be the transformation $z \mapsto z/(4z + 1)$ and G_1 be the group generated by g_0 and g_1. Then G_1 is a Fuchsian group without elliptic elements and $g((-1 + i)/4) = (1 + i)/4$.

Theorem 3.10.6 implies now immediately the following result:

Theorem 3.10.7 *Let X be a hyperbolic Riemann surface with punctures. Each puncture of X has a horocyclic neighborhood of area 4. The inner boundary curve of this horocycle has length 4.*

Observe that the horocyclic neighborhoods of Theorem 3.10.6 at disjoint punctures are not necessarily disjoint. An example of this situation is provided by the group G_1. Area 4 horocycle associated to the element g_0 is the euclidean disk of radius $\frac{1}{2}$ with center at $\frac{i}{2}$. Area 4 horocycle associated to the element $g_1(z) = z + 1$ is the half-plane $\operatorname{Im} z > \frac{1}{4}$. They overlap.

Let g be any parabolic Möbius transformation mapping the upper half-plane onto itself. Then we may always assume that g is conjugate either to the transformation $z \to z + 1$ or to $z \to z - 1$. If g is conjugate to $z \to z + 1$, then g^{-1} is conjugate to $z \to z - 1$. The orientation reversing Möbius transformation $z \mapsto -\overline{z}$ conjugates $z \mapsto z + 1$ to $z \mapsto z - 1$.

Assume now that G is any Fuchsian group acting in the upper half-plane U and let g be a primitive parabolic element in G. The above observation together with Lemma 3.10.2 implies that there is always a hyperbolic disk (or a half-plane) $D_g \in U$ such that ∂D_g is tangent to ∂U at the fixed-point of g and if $h(D_g) \cap D_g \neq \emptyset$, then $h \in \langle g \rangle$.

In accordance with Definition 3.10.2 we say that the disk D_g is a *horocycle* of the parabolic transformation g.

3.11 Nielsen's criterium for discontinuity

Theorem 3.10.6 allows us to give a fairly general criterium that guarantees the discontinuity of a Möbius group acting in the upper half-plane.

Theorem 3.11.1 *Assume that G is a group of Möbius transformations mapping the upper half-plane U onto itself and containing the translation*

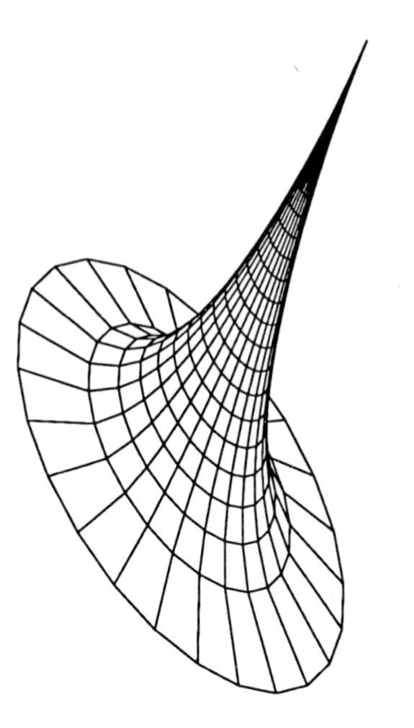

Figure 3.8: A horocyclic neighborhood of a puncture on a hyperbolic Riemann surface. This surface is the surface of revolution of the tracktrix curve which is characterized by the property that its tangent line meets the x–axis at unit distance from the point of tangency. This surface has curvature -1.

$g_1(z) = z + 1$. *If G does not contain elliptic elements, then G is either properly discontinuous or contains only elements fixing ∞.*

Proof. Suppose that G contains elements not fixing the infinity. In view of Theorem 3.5.4 (on page 84) it suffices to show that the regular set $\Omega(G)$ of G contains a point in the upper half-plane.

If $g \in G$ does not fix the infinity and $g_\omega(z) = z + \omega$ is in G, then, by Lemma 3.10.5,

$$\omega \geq 4r_g. \tag{3.23}$$

From (3.23) it follows that the subgroup G_∞ of the *translations* of G is cyclic.

We show next that the group G does not contain any hyperbolic elements fixing the infinity. Assume, on the contrary, that $h \in G$ is hyperbolic and $h(\infty) = \infty$. Then we may suppose that $h(z) = kz$ for some $k > 1$.

Observe that r_g is not changed if g is conjugated by a translation. Now, $h^{-n} \circ g_\omega \circ h^n \in G_\infty$ and

$$(h^{-n} \circ g_\omega \circ h^n)(z) = z + k^{-n}.$$

This contradicts (3.23).

Let $g_\omega(z) = z + \omega$, $\omega > 0$, be a generator for G_∞. Choose $z_0 \in U$ such that $\operatorname{Im} z_0 > \frac{\omega}{2} + \frac{1}{4}$. By Theorem 3.10.6, the disk $D(z_0, \omega/2) = \{z \mid |z - z_0| < \omega/2\}$ does not contain points $z, z \neq z_0$, that are equivalent to z_0 under G. We conclude that $z_0 \in \Omega(G)$.

Theorem 3.11.1 is a special case of the following more general result.

Theorem 3.11.2 *Assume that G is a group of Möbius transformations mapping the upper half-plane onto itself. If the group G does not contain elliptic elements, then it is discontinuous.*

In view of Theorem 3.11.1 we have to shown only that purely hyperbolic Möbius groups fixing the upper half-plane is discontinuous. Proof for this can be found in Appendix B, Theorem A.8.5 (on page 240).

The above result is a special case of a more general result stating that *a Möbius group mapping the upper half-plane onto itself is discontinuous if it does not contain infinitesimal elliptic elements.* This has been first shown by C. L. Siegel in [85], where Siegel calls this theorem 'a result of Jakob Nielsen'.

3.12 Classification of Fuchsian groups

Let D_g be a horocycle associated to a parabolic Möbius transformation g. It is immediate that $D_g/\langle g\rangle$ is conformally equivalent to the punctured unit disk $D^* = \{z \mid 0 < |z| < 1\}$. Assuming that g is a primitive element, we conclude, by Lemma 3.10.2, that $D_g/\langle g\rangle$ is conformally homeomorphic to an open subset of the Riemann surface U/G. Using this non–compact subset of U/G it is easy to construct an open covering of U/G which does not have a finite subcovering of U/G. This implies that U/G is not compact. We have, therefore, the following result.

Theorem 3.12.1 *Let G be a Fuchsian group acting in the upper half-plane U. If U/G is a compact Riemann surface, then the group G does not contain parabolic elements.*

By Theorem 3.12.1 we conclude that all non–identity elements of a Fuchsian group corresponding to a compact Riemann surface are hyperbolic Möbius transformations.

Let now γ be a closed curve on a Riemann surface U/G. Let $g = g_\gamma$ be a Möbius transformation corresponding to the homotopy class of the curve γ as explained in Lemma 3.6.2 and in Lemma 3.6.3.

Recall that for a hyperbolic Möbius transformation $\inf_{z\in U} d_U(z, g(z))$ is obtained for any $z \in ax(g)$. Here $ax(g)$ denotes the axis of the hyperbolic transformation g and d_U is the hyperbolic metric of the upper half–plane U. We have, furthermore, $\inf_{z\in U} d_U(z, g(z)) = \log k(g)$, where $k(g) > 1$ is the multiplier of g.

We conclude therefore that if the transformation g_γ is hyperbolic, then the homotopy class of the curve γ contains a geodesic curve which is the projection of the geodesic arc from z to $g_\gamma(z)$ for any point $z \in ax(g_\gamma)$.

Furthermore we conclude that if g_γ is parabolic, then the homotopy class of the curve γ does not contain any geodesic curves. In this case we say that γ is a curve *going around a puncture of U/G.*

Let G be a Fuchsian group acting in the upper half–plane.

Definition 3.12.1 *Let $G(z) = \{g(z) \mid g \in G\}$ denote the orbit of a pont z under the action of G. A point $x \in \mathbf{R} \cup \{\infty\}$ is a* limit point *of G if there exists a point $z \in U$ such that x belongs to the closure of $G(z)$. The set $L(G)$ of G.*

Definition 3.12.2 *A Fuchsian group G acting in the upper half-plane is said to be of the* first kind *if $L(G) = \mathbf{R} \cup \{\infty\}$. Groups that are not of the first kind are said to be of the* second kind.

This terminology is standard today. The reader should observe, however, that, in old literature, groups of the first kind were sometimes called groups of the second kind (and vica versa). For instance in a paper of Burnside ([18]) this terminology was used but the meaning was the contrary.

Lemma 3.12.2 *Assume that the Riemann surface U/G is compact. Then the Fuchsian group G is of the first kind.*

Proof. This argument is due to Pekka Tukia. Let $r \in \mathbf{R}$ be an arbitrary point in the real line. It suffices to show that the closer of the orbit of the point $i \in U$ contains the whole real line.

Let $\epsilon > 0$ be arbitrary. Let

$$U_\epsilon(r) = \{z \in \mathbf{C} \mid |z - r| < \epsilon\}$$

denote the euclidean disk of radius ϵ and center at $r \in \mathbf{R}$. By the properties of the hyperbolic metric, $U_\epsilon(r) \cap U$ contains hyperbolic disks D_R of arbitrarily large radius R. Now the image of any hyperbolic disk of radius

$$R > \text{diameter of} \, U/G = \sup\{d_{U/G}(p,q) \mid p,\, q \in U/G\}$$

under the projection $\pi : U \to U/G$ necessarily covers the whole Riemann surface U/G. Therefore any such disk contains a point z such that $\pi(i) = \pi(z)$. This point z belongs to the orbit $\{g(i) \mid g \in G\}$ of i. We conclude, therefore, that $r \in \mathbf{R}$ belongs to the closure of the orbit of i.

Repeating this argument we could prove the above observation stating that i can be replaced by an arbitrary point z in the upper half–plane. Likewise one can show that the limit set of a Fuchsian group is the closer of the set fixed–points of non–identity elements of that group.

3.13 Short closed curves

Theorem 3.12.1 allows us to apply considerations of Chapter 1 to estimate lengths of intersecting closed geodesic curves on a hyperbolic Riemann surface. We will next show that if two closed geodesic curves intersect, then they cannot both be short.

Lemma 3.13.1 *There exists a universal constant η, $\eta > 0$, such that for any compact Riemann surface X (which may have boundary components) the following is true: Let α and β be closed geodesic curves on W with lengths $< \eta$. Then either $\alpha = \beta$ (as set of points) or the curves α and β do not intersect.*

Proof. This result follows directly from inequality (1.25) in Corollary 1.5.5 on page 33. Assume first that α and β are closed and intersecting geodesic curves such that $\alpha \neq \beta$ (as sets of points). Write $W = U/G$ for a Fuchsian group G. Let g and h be a hyperbolic transformations, g covering α and h covering β (cf. Definition 3.6.1). Provided that g and h are obtained using the same isomorphism i of Lemma 3.6.3, g and h are now hyperbolic Möbius transformations with intersecting axes.

Applying the arguments of Theorem 3.12.1 we can conclude that the non–identity elements of the group G are hyperbolic Möbius transformations. This implies that the commutator $c = [g, h]$ of the hyperbolic transformations g and h is hyperbolic as well. Therefore the assumptions of Corollary 1.5.5 are satisfied.

Let ℓ_α and ℓ_β be the lengths of the geodesic curves α and β, respectively. Then the multiplier k_1 of g satisfies $k_1 = e^{\ell_\alpha}$ and that of h is $k_2 = e^{\ell_\beta}$.

Inequality (1.25) gives then the following inequality for the lengths of the curves α and β :

$$\frac{4e^{\ell_\beta}}{(e^{\ell_\beta} - 1)^2} < \frac{(e^{\ell_\alpha} - 1)^2}{4e^{\ell_\alpha}}. \tag{3.24}$$

Constant η is then the positive solution of the equation

$$\frac{4e^{\eta}}{(e^{\eta} - 1)^2} = \frac{(e^{\eta} - 1)^2}{4e^{\eta}}. \tag{3.25}$$

A numerical estimate for the positive solution of equation (3.25) is $\eta \approx 1.33254$.

Another inequality of the same type as inequality (3.24) is given in [1, Lemma 1 on page 94]. It is not quite as strong as the one presented here. It yields the numeric estimate 1.01859 for the constant η.

It is interesting to observe that the above inequality was obtained using only the facts that the Möbius transformations g, h and their commutator $c = [g, h]$ are hyperbolic. The discontinuity of the action of the Fuchsian group G does not play any role here.

3.14 Collars

Definition 3.14.1 *Let $A \subset X$ be a non–empty subset of a (hyperbolic) Riemann surface X and let ϵ be a positive number. The* ϵ–distance neighborhood *of A is the set*

$$N_\epsilon(A) = \{p \in X \mid d_X(p, A) < \epsilon\}.$$

Let X now be a hyperbolic Riemann surface, $p \in X$ a point and α a simple closed geodesic curve on X. For an $\epsilon > 0$ we say that $N_\epsilon(\{p\})$ is a *disk* if $N_\epsilon(\{p\})$ is homeomorphic to a disk in the complex plane.

Likewise we say that $N_\epsilon(\{p\})$ is a *collar* if it is homeomorphic to an annulus in the complex plane. We say further that a collar $N_\epsilon(\alpha)$ has *width* ϵ.

Let α be a simple closed geodesic curve *in the interior* of X. A collar $N_\epsilon(\{p\})$ of area 2μ is called a *μ–collar at α*. If α is a boundary component of X, then a collar $N_\epsilon(\alpha)$ of area μ is called a μ–collar. The area and the width of a collar at a geodesic curve of length ℓ are related to each other in the following way.

Lemma 3.14.1 *Assume that α is a simple closed geodesic curve in the interior of a hyperbolic Riemann surface X and that $N_\epsilon(\alpha)$ is a μ–collar at α. Then we have*

$$\mu = \ell_\alpha \sinh \epsilon$$

where ℓ_α denotes the length of α on X.

The proof of this result is an elementary computation in hyperbolic geometry. We will leave the details to the reader. Arguments are similar to the ones of the proof of Lemma 3.14.7.

Next we use considerations of Chapter 1 to estimate the distance $d_X(\alpha, \beta)$ between closed geodesic curves α and β on a Riemann surface $X = D/G$.

By the considerations based on Theorem 3.12.1 we can associate to any closed geodesic curves α and β two hyperbolic Möbius transformations $g \in G$ and $h \in G$ such that the axis of g and h project onto α and β, respectively. Furthermore, if $k_1 = k(g)$ and $k_2 = k(h)$ are the multipliers of g and h, then $\ell_\alpha = \log k_1$ and $\ell_\beta = \log k_2$.

Let $d_X(\alpha, \beta) = \log k$. If α and β do not intersect each other, then we can deduce, by Theorem 1.5.7, that

$$k > \frac{f(\sqrt{k_1 k_2}) + f(\sqrt{k_1/k_2})}{f(\sqrt{k_1 k_2}) - f(\sqrt{k_1/k_2})}. \tag{3.26}$$

where $f(t) = \sqrt{t} + 1/\sqrt{t}$ for $t > 0$.

Estimate (3.26) gives us immediately an estimate for the distance $d_X(\alpha, \beta)$ in the following way. The right hand side of (3.26) can be rewritten so that one has

$$k > \frac{\left(\sqrt[4]{k_1} + \frac{1}{\sqrt[4]{k_1}}\right)\left(\sqrt[4]{k_2} + \frac{1}{\sqrt[4]{k_2}}\right)}{\left(\sqrt[4]{k_1} - \frac{1}{\sqrt[4]{k_1}}\right)\left(\sqrt[4]{k_2} - \frac{1}{\sqrt[4]{k_2}}\right)}. \tag{3.27}$$

Estimate 3.27 implies the following result.

Lemma 3.14.2 *Let α and β be closed geodesic curves of length ℓ_α and ℓ_β, respectively, on a hyperbolic Riemann surface. Then either $d_X(\alpha,\beta) = 0$ (in which case α and β intersect) or*

$$d_X(\alpha,\beta) > \log\coth(\ell_\alpha/4) + \log\coth(\ell_\beta/4) \tag{3.28}$$

Estimate (3.28) means that short non–intersecting curves on any hyperbolic Riemann surface are rather far apart from each other.

Estimate (3.28) has important applications for us.

Let

$$\delta(\alpha) = \inf\{d_X(\alpha,\beta) \mid \beta \text{ a closed geodesic curve } \beta\cap\alpha = \emptyset\}. \tag{3.29}$$

For a fixed value of ℓ_α, the right hand side of (3.28) is a decreasing function of ℓ_β tending to the limit

$$\log\coth(\ell_\alpha/4)$$

as $\ell_\beta \to \infty$.

This shows the following result:

Lemma 3.14.3 *For a simple closed geodesic curve α,*

$$\delta(\alpha) \geq \log\coth \ell_\alpha/4.$$

We will now use the above considerations to estimate the size of the maximal collar at a simple closed geodesic curve. To that end observe the following result.

Lemma 3.14.4 *A simple closed geodesic curve α on a hyperbolic Riemann surface X has always a collar $N_{\delta(\alpha)}(\alpha)$ of width $\delta(\alpha)$.*

Proof. It is clearly enough to show the following:

Lemma 3.14.5 *If the closure $\overline{N_\epsilon(\alpha)}$ of a ϵ–distance neighborhood $N_\epsilon(\alpha)$ of a simple closed geodesic curve α on a hyperbolic Riemann surface X is not homeomorphic to $\alpha\times[-\epsilon,\epsilon]$, then there exists a closed geodesic curve β that does not intersect α such that $\beta\cap\overline{N_\epsilon(\alpha)}\neq\emptyset$.*

Proof. It is enough to consider the case where $N_\epsilon(\alpha) \approx \alpha\times(-\epsilon,\epsilon)$ and $\overline{N_\epsilon(\alpha)} \not\approx \alpha\times[-\epsilon,\epsilon]$. Then $N_\epsilon(\alpha)$ has either one or two boundary components. Let us assume that $N_\epsilon(\alpha)$ has two boundary components. The case of only one boundary component can be treated in the same way as the present case and will be left to the reader. Figure 3.9 illustrates this case. Let α' and α'' denote the boundary components of $N_\epsilon(\alpha)$. At least one of them,

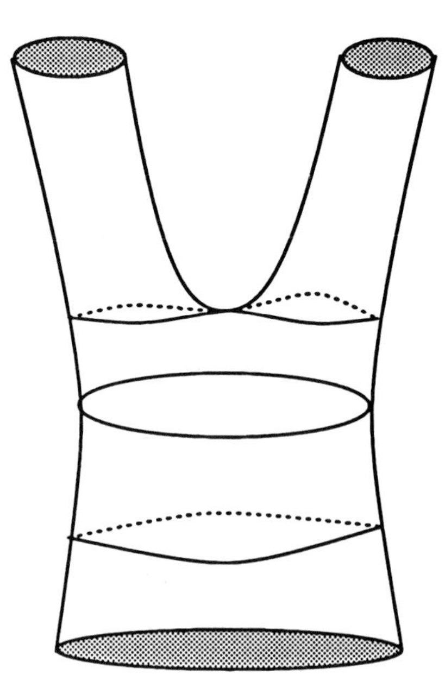

Figure 3.9: $N_\epsilon(\alpha)$ has two boundary components.

say α', is not anymore a simple closed curve. Then α' can be expressed as a finite union of simple closed curves $\gamma_1, \ldots, \gamma_n$, $n \geq 2$, such that γ_i and γ_{i+1} intersect at one point p_i. Consider the curves γ_1 and γ_2 intersecting at the point p_1. Assume that both curves have the positive boundary orientation (as parts of the boundary of $N_\epsilon(\alpha)$). Let γ be the geodesic curve freely homotopic to $\gamma_1\gamma_2^{-1}$. Now γ is a geodesic curve which is not homotopic to α. γ is furthermore homotopic to a closed curve that does not intersect α. We deduce, therefore, that $\alpha \cap \gamma = \emptyset$. For topological reasons it is, on the other hand, clear that $\gamma \cap \overline{N_\epsilon(\alpha)} \neq \emptyset$.□

Lemmata 3.14.4 and 3.14.3 imply now the following result:

Theorem 3.14.6 *Let α be a simple closed geodesic curve of length ℓ_α on a hyperbolic Riemann surface X. The curve α has always a μ–collar $N_\epsilon(\alpha)$ for $\mu = \mu(\ell_\alpha) = \ell_\alpha \sinh(\log\coth(\ell_\alpha/4))$. The width of this collar is $\epsilon = \log\coth \ell_\alpha/4$.*

Observe that the above function μ is a positive decreasing function and that

$$\lim_{\ell \to 0+} \mu(\ell) = 2. \tag{3.30}$$

In order to apply the above results it is necessary to take a closer look at the geometry of collars. We are mainly interested in collars at boundary components of hyperbolic Riemann surfaces. We will, therefore, assume that α is a boundary geodesic of a hyperbolic Riemann surface X. What is said below is, nevertheless, true also for general simple closed geodesic curves. Let $\epsilon > 0$. Assume that the ϵ–distance neighborhood $N_\epsilon(\alpha)$ for a boundary component α is a collar at α.

The collar $N_\epsilon(\alpha)$ has itself two boundary components α and α^*. The latter is called *the inner boundary component,* while α is *the outer boundary component.* Let ℓ be the length of α and ℓ^* that of α^*.

Lemma 3.14.7 *The length of the inner boundary component α^* of a μ–collar $N_\epsilon(\alpha)$ at α is*

$$\ell^* = \sqrt{\ell^2 + \mu^2}. \tag{3.31}$$

Any simple closed curve that is freely homotopic to α and lies outside of this collar has length at least ℓ^*.

Proof ([13, Lemma 4, page 89]). Let W^S be the Schottky double of W. Express $W = U/G$, where G is a Fuchsian group acting in the upper half–plane U. Let $Q \in \alpha$ be a base point and $i : \pi_1(W, Q) \to G$ the isomorphism defined on page 87. The choice of the point Q here is immaterial. Giving an orientation to the curve α we can now interpret α as a curve defining a point of the fundamental group $\pi_1(W, Q)$. Here the orientation chosen for α does not play any role. Important is that, via the isomorphism i : $\pi_1(W, Q) \to G$ we may associate a Möbius transformation g to the curve α. Choosing a different orientation for α would only mean that we associate g^{-1} to α rather than g. The Möbius transformation g is hyperbolic by Theorem 3.12.1. Then, by the definition of the isomorphism i, the axis of g projects onto a simple closed geodesic curve on W^S which is homotopic to the curve α. But α itself is also a geodesic curve. In each homotopy class of closed curves on W^S there is a unique geodesic curve.

We conclude that the axis of g projects onto the curve α under the projection

$$U \to W^S = U/G. \tag{3.32}$$

By conjugation we may assume that $g(z) = e^\ell z$. Then the positive imaginary axis is the axis of g. Considerations on page 24 (see Fig. 1.1) imply now that, for some θ_0, $N_\epsilon(\alpha)$ is the image of a region

$$\{z \in U | 1 \leq |z| < e^\ell, \ |\arg z - \frac{\pi}{2}| < \theta_0, \ \mathrm{Re}\, z \geq 0\}$$

under the projection (3.32). The projection is, furthermore, injective on this region.

Use the polar coordinates r, θ in U setting $z = ire^{i\theta}$, $-\pi/2 < \theta < \pi/2$. The hyperbolic metric of U becomes $((dr^2/r^2) + d\theta^2)/\cos^2\theta$. Direct computation gives then

$$\mu = \ell \tan \theta_0. \tag{3.33}$$

The inner boundary curve of the μ–collar $N_\epsilon(\alpha)$ at α is now the image of the segment

$$\{(r,\theta) | 1 \leq r \leq e^\ell,\, \theta = \theta_0\}$$

under the projection (3.32). An easy calculation shows that this segment has length

$$\ell^* = \frac{\ell}{\cos\theta_0}.$$

These expressions obtained for ℓ^* and for μ imply now (3.31).□

Lemma 3.14.8 *A μ–collar at a boundary component of length ℓ contains a γ–collar for every γ, $0 < \gamma < \mu$. Every point of the inner boundary component the former lies at the distance*

$$\log \frac{\mu + \sqrt{\mu^2 + \ell^2}}{\gamma + \sqrt{\gamma^2 + \ell^2}} \tag{3.34}$$

from the inner boundary component of the latter.

Proof. It is obvious that a μ–collar contains every γ–collar for $\gamma < \mu$. The distance between the inner boundary components of such collars is simply the difference in the widths of these collars. So the proof amounts in computing the width of a μ–collar.

Express our μ–collar as in Lemma 3.14.7. Let $w(\mu)$ denote the width of the collar. It can be computed in terms of polar coordinates. We obtain:

$$w(\mu) = \int_{\theta_0}^{\pi/2} \frac{r d\theta}{r\cos\theta} = \log\left(\frac{1+\sin\theta_0}{\cos\theta_0}\right). \tag{3.35}$$

Equation (3.33) in conjunction with equation (3.35) implies now

$$w(\mu) = \log\left(\frac{\mu}{\ell} + \sqrt{1 + \left(\frac{\mu}{\ell}\right)^2}\right). \tag{3.36}$$

This expression for the width implies now equation (3.34).

Now we can compute a corollary to Theorem 3.14.6 by the above results:

Corollary 3.14.9 *Every boundary geodesic α on a Riemann surface X has a collar whose inner boundary component is a simple closed curve of length > 2.*

Proof. Let α be a boundary geodesic of length ℓ. Consider the collar of Theorem 3.14.6.

By formula (3.31) the inner boundary curve of this collar has the length

$$\ell^* = \ell^*(\ell) = \sqrt{\ell^2 + (\mu(\ell))^2} \tag{3.37}$$

where $\mu(\ell) = \ell \sinh(\log \coth \ell/4)$ is the function given in Theorem 3.14.6. Elementary analysis shows that $\ell^*(\ell) > 2$ for all $\ell > 0$.

Theorem 3.14.6 is not the best possible result in this direction. Explicit calculations about the hyperbolic geometry of pairs of pants and hyperbolic polygons show the following result.

Lemma 3.14.10 *Define the function $\mu(\ell)$ setting*

$$\mu(\ell) = \frac{\ell}{\sinh(\ell/2)}.$$

Every simple closed geodesic curve α on a hyperbolic Riemann surface X has a $\mu(\ell)$–collar for every $\ell \leq \ell_\alpha$, where ℓ_α is the length of α. Furthermore, two such collars at disjoint simple closed geodesic curves are disjoint as well.

This result is known as the Keen Collar Lemma. We will not prove it here because the slightly weaker version of the result, Theorem 3.14.6, is sufficient for our purposes. A clear and elementary proof of this result can be found in the forthcoming monograph of Peter Buser ([22]). For other proofs of related (and weaker) versions of this result see e.g. [44], [36] or [1, Pages 95–96].

Above results give us information about the location of short geodesics on a hyperbolic Riemann surface. We conclude these deliberations by considering the set of the lengths of *all* closed geodesic curves.

3.15 Length spectrum

Definition 3.15.1 *Let U/G be a Riemann surface. The set of the lengths of closed geodesics on U/G is called the* length spectrum *of the Riemann surface U/G. It is denoted by $\mathcal{L}(U/G)$.*

The number of homotopy classes of closed curves on a compact Riemann surface is countable. Since each homotopy class contains only one geodesic curve, also the length spectrum of a (compact) Riemann surface is (at most) a countable set.

Theorem 3.15.1 *The length spectrum $\mathcal{L}(U/G) = \{\ell_1, \ell_2, \ldots\}$ of a compact Riemann surface U/G is a discrete subset of* $\mathbf{R}$.

Proof. Assume the contrary. Then there exists a constant M and infinitely many closed geodesic curves γ_i, $i = 1, 2, \ldots$ such that $\ell_i = \ell(\gamma_i) < M$, where $\ell(\gamma_i)$ denotes the length of the hyperbolic geodesic γ_i.

Let $\pi : U \to U/G$ denote the projection. Choose a point $p \in U/G$ and points $p_i \in \gamma_i$ such that $d(p, p_i) = d(p, \gamma_i)$, where d denotes the hyperbolic distance.

Choose a point $\tilde{p} \in \pi^{-1}(p)$ and points $\tilde{p}_i \in \pi^{-1}(p_i)$ in such a way that $d_U(\tilde{p}, \tilde{p}_i) = d(p, p_i)$. Here d_U denotes the hyperbolic metric of the upper half–plane U.

Assume that $g_i \in G$ covers the homotopy class of the curve γ_i (as explained in Lemma 3.6.3) and that $\tilde{p}_i$ lies on the axis of the transformation g_i. Then

$$\begin{aligned} d_U(\tilde{p}, g_i(\tilde{p})) &\leq d_U(\tilde{p}, \tilde{p}_i) + \ell(\gamma_i) + d_U(g_i(\tilde{p}), g_i(\tilde{p}_i)) \\ &= 2d_U(\tilde{p}, \tilde{p}_i) + \ell(\gamma_i) \\ &= 2d(p, p_i) + \ell(\gamma_i). \end{aligned}$$

Since G acts discontinuously and since the set γ_i contains infinitely many different curves γ_i we conclude that $d_U(\tilde{p}, \tilde{p}_i) \to \infty$. This means that $d(p, p_i) \to \infty$ on U/G. But that is not possible since U/G is assumed to be compact.

Corollary 3.15.2 *Among closed geodesic curves on a (compact) Riemann surface there are curves with minimal length.*

This corollary follows immediately from Theorem 3.15.1. It leads to the following interesting question:

> Let $\epsilon(X)$ denote the length of shortest closed geodesic curve on the Riemann surface X. Define
>
> $$L^g = \sup \epsilon(X)$$
>
> where supremum is taken over all Riemann surfaces of genus g. Estimate the number L^g for each g.

Observation 3.15.3 *Theorem 3.15.1 can immediately be generalized to the case of a Riemann surface that is obtained from a compact Riemann surface U/G by deleting finitely many points. Such Riemann surfaces can be Uniformized by a Fuchsian group G' which contains parabolic elements. These parabolic elements correspond to the punctures of U/G', i.e., to points that were deleted from U/G in order to get U/G'. The above argument can be applied also in this more general case. Details are left to the reader.*

We conclude this section by the following remarks. Keeping the differential geometric aspect in mind, we call isomorphic Riemann surfaces *isometric*. Riemann surfaces having the same length spectrum are, on the other hand, called *isospectric*. If X and Y are isometric Riemann surfaces, then clearly they have the same length spectrum. The converse is not true: Marie–France Vigneras has first constructed examples ([95]) of hyperbolic Riemann surfaces that have the same length spectrum but which are not isomorphic to each other. These examples of Vigneras had a very large genus. Peter Buser has subsequently shown ([21], see also [20]) such examples for genera $g = 5$ and $g \geq 7$. R. Brooks ([15]) has generalized this approach of Buser and constructed examples of isospectric and non–isometric Riemann surfaces of genus 4 and 6.

In the case of one–holed torus and Riemann surfaces of genus two, the situation is different. Peter Buser and Klaus–Dieter Semmler have shown (cf. [19]) that in these cases isospectric Riemann surfaces are also isometric. It is not known whether isospectric Riemann surfaces of genus three are necessarily isometric.

The above remarks lead to the following question: How many non–isometric Riemann surfaces of genus g can have the same length spectrum? H. P. McKean ([62] and [63]) has first shown that this number is finite. Peter Buser ([20]) has then shown that at most $507g^3$ non–isometric Riemann surfaces of genus g can have the same length spectrum.

3.16 Pants decompositions of compact surfaces

Let Σ be a compact and oriented topological surface of genus $g > 1$. Then $\chi(\Sigma) < 0$.

Definition 3.16.1 A standard pair of pants P *is the domain*

$$P = \{z \in \mathbf{C} | |z| \leq 2, |z-1| \geq \frac{1}{2}, |z+1| \geq \frac{1}{2}\},$$

whose three boundary components are order as follows: $\gamma_1 = \{z \in \mathbf{C} | |z| = 1,$ $\gamma_2 = \{z \in \mathbf{C} | |z-1| = \frac{1}{2}\}$ *and* $\gamma_3 = \{z \in \mathbf{C} | |z+1| = \frac{1}{2}\}$.

For technical convenience we have here assumed that the boundary curves of a standard pair of pants form an ordered set. Then we may speak of *the first*, *the second* and of *the third* boundary component.

A standard pair of pants is a triply connected domain which is clearly homeomorphic to any usual pair of pants.

Definition 3.16.2 *A topological Hausdorff space P' together with a homeomorphism $h : P' \to P$ onto the standard pair of pants is called* a pair of pants.

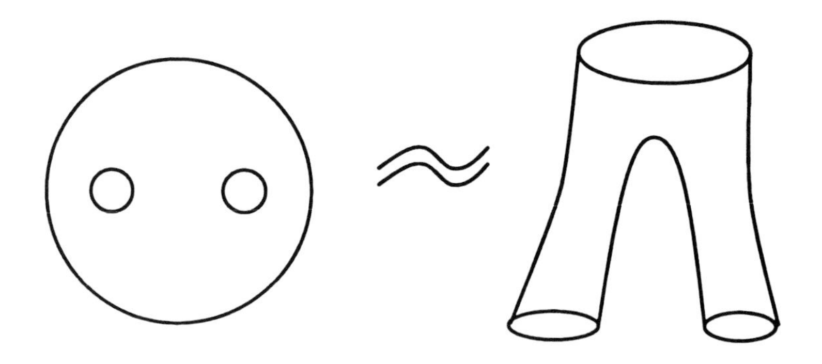

Figure 3.10: A standard pair of pants is homeomorphic to a usual pair of pants.

Each pair of pants is a surface with three boundary components. The boundary components of the standard pair of pants form an ordered set. Via the homeomorphism $h : P' \to P$ the set of boundary components of P' gets also an order. We may, therefore, always speak of the first, second and third boundary component of a pair of pants.

Pairs of pants are building blocks for surfaces. It is easy to see that one can build all compact surfaces with boundary components starting from a number of pairs of pants $P_1, \ldots, P_n$ and gluing them together in some suitable way along the boundary components.

Observe that it is also possible to build non–orientable surface from pairs of pants (that are orientable themselves). That is done by changing — in a suitable fashion — the orientations in which the boundary components of the pairs of pants are glued together.

Elementary computation gives the value $\chi(P) = -1$ for the Euler characteristic of a pair of pants. If A and B are two subsets of the same manifold, then — provided that the respective Euler characteristics are finite — $\chi(A \cup B) = \chi(A) + \chi(B) - \chi(A \cap B)$.

Let X be a manifold obtained by gluing two pairs of pants P_1 and P_2 together along their first boundary components. Then $P_1 \cap P_2$ is a circle and $\chi(P_1 \cap P_2) = 0$. We conclude that $\chi(X) = \chi(P_1) + \chi(P_2)$. This arguments can immediately be generalized to obtain the following result.

Theorem 3.16.1 *Each compact surface Σ, with $\chi(\Sigma) < 0$, can be composed by gluing the boundary curves of $-\chi(\Sigma)$ pairs of pants together in a suitable fashion.*

By Theorem 3.16.1 we may always decomposed a surface Σ, $\chi(\Sigma) < 0$, into $-\chi(\Sigma)$ pairs of pants. Such a decomposition

$$\mathcal{P} = \{P_1, P_2, \ldots, P_{-\chi(\Sigma)}\}$$

is understood to be *ordered*. So we can always — when referring to a pants decomposition — speak of the first pairs of pants, the second etc.

Let $\mathcal{P}$ be a decomposition of Σ into pairs of pants. The boundary curves of the respective pairs of pants are simple closed curves $\alpha_1, \alpha_2, \ldots, \alpha_m$ on the surface Σ. We call these curves *the decomposing curves.* We assume that the decomposing curves are given *some* orientation. Therefore we may speak of the *positive* direction of a decomposing curve.

We conclude that:

- Decomposing curves are simple closed curves on Σ.
- The decomposing curves do not intersect.
- No pair of the decomposing curves are homotopic to each other (or to each other's inverse).

Any set of decomposing curves is a maximal set with these properties. Recall that we have agreed that any decomposition $\mathcal{P}$ of a surface Σ into pairs of pants is an *ordered* collection of pairs of pants. Also the boundary curves of any pair of pants is an ordered collection of simple closed curves as well. Therefore the set of decomposing curves of a pants decomposition $\mathcal{P}$ is also an ordered set, the order being defined by all the other orderings.

Next consider the case of symmetric surfaces. Recall Definition 3.4.1: a symmetric surface is a (topological) surface Σ together with an orientation reversing involution $\sigma : \Sigma \to \Sigma$.

For certain applications it is important to study pants decompositions of symmetric surfaces. The decompositions themselves should be symmetric as well. We start with some technicalities.

Let Σ be a surface with boundary components and $\sigma : \Sigma \to \Sigma$ an orientation reversing involution. A pair (α_1, α_2) of simple closed curves α_1 and α_2 on Σ is called *a σ–pair* if $\sigma(\alpha_1) = \alpha_2$ and if either $\alpha_1 = \alpha_2$ or the curves are disjoint. A σ–pair (α_1, α_2) is called *essential* if α_1 is not freely homotopic to any boundary component of Σ. Observe that this condition implies that for an essential σ–pair (α_1, α_2) neither one of the curves α_j is freely homotopic to a boundary component.

Assume that X is such a complex structure on Σ that the mapping $\sigma : (\Sigma, X) \to (\Sigma, X)$ is antiholomorphic. We equip the Riemann surface $X = (\Sigma, X)$ with the usual intrinsic hyperbolic metric.

A σ–pair (α_1, α_2) on the Riemann surface X is *geodesic* if both curves α_j, $j = 1, 2$, are geodesic curves in the intrinsic hyperbolic metric of (Σ, X). Since $\sigma : X \to X$ is now an isometry of the hyperbolic metric, both curves α_1 and α_2 have the same length, which is referred to as the *length* of the σ–pair (α_1, α_2).

Next sections are devoted to showing that there always exists a σ–invariant set of $3g - 3$ disjoint and simple closed geodesic curves on X in such a way that the length of the longest one is bounded by $21g$. This set of curves then determines a σ–invariant, i.e., a symmetric decomposition of X into pairs of pants.

We will construct this decomposition in an inductive way. We first choose a σ–pair that is as short as possible. It turns out that it can be always chosen in such a way that its length is less than $7g + 1$. This is shown below.

We cut the surface open along this σ–pair to obtain a Riemann surface W_0 with boundary components. This Riemann surface needs not be connected anymore. We start in an inductive way to cut the Riemann surface W_0 into smaller parts in a way that is, at each step, compatible with the action of σ and uses as short curves as possible.

Here we are considering symmetric surfaces. The same argument applies to all compact Riemann surfaces. A compact Riemann surface of genus g can always be decomposed into pairs of pants by simple closed geodesic curves of length $< 21g$. That follows from the present arguments. We consider here only the more complicated case of symmetric surfaces. The arguments that we present in the subsequent sections are taken from [23].

3.17 Shortest curves on a hyperbolic Riemann surface with a symmetry

We assume now that $X = (\Sigma, X)$ is a fixed compact Riemann surface of genus g, $g > 1$, and $\sigma : (\Sigma, X) \to (\Sigma, X)$ an antiholomorphic involution.

Lemma 3.17.1 *Let ω be a piecewise geodesic closed curve with at most finitely many double points on a Riemann surface X. If ω has at most two points where it is not smooth, then ω is not homotopic to a point.*

Proof. Assume that ω is homotopic to a point and consider the universal covering $\pi : U \to X$. If ω is homotopic to a point, then there is a lift $\tilde{\omega}$ of ω to U which is a closed curve and a union of at most two geodesic arcs in U. But this is clearly impossible.

Lemma 3.17.2 *Let α be a closed geodesic curve on X with minimal length. Then α is simple and*

$$\ell_\alpha \leq 2 \operatorname{arcosh}(2g-1) < 2\log(4g-2). \tag{3.38}$$

Proof. It is clear that a curve with minimal length is simple. Let $p \in \alpha$. Consider the sets

$$N_r(\{p\}) = \{q \in X \mid \operatorname{dist}(p,q) < r\}.$$

For $0 < r \leq \frac{1}{2}\ell_\alpha$ the set $N_r(\{p\})$ is isometric to a hyperbolic disk in the upper half–plane U and has area (cf. Theorem A.3.1)

$$\text{area } N_r(\{p\}) = 2\pi(\cosh r - 1).$$

On the other hand, $N_r(\{p\}) \subset X$ and

$$\text{area } N_r(\{p\}) \leq \text{area } X = 4\pi(g-1)$$

where g is the genus of X. Taking $r = \frac{1}{2}\ell_\alpha$ we get

$$2\pi\left(\cosh(\frac{1}{2}\ell_\alpha) - 1\right) \leq 4\pi(g-1).$$

This yields inequality (3.38).

Lemma 3.17.3 *Let α be a simple closed curve of minimal length on X and $\sigma : X \to X$ an antiholomorphic involution. Assume that $\sigma(\alpha) \cap \alpha \neq \emptyset$. Then either $\sigma(\alpha) = \alpha$ or $\sigma(\alpha)$ intersects α at exactly one point.*

Proof. If neither is the case, then α intersects $\sigma(\alpha)$ at some finite number of points $p_1, \ldots, p_n$, $n > 1$. Let p_1 and p_2 be two consecutive intersection points. Let α' be an arc of the geodesic curve α with end–points p_1 and p_2 and which has length $\leq \frac{1}{2}\ell_\alpha$. Let α'' be similar arc on the geodesic curve $\sigma(\alpha)$ also with end–points p_1 and p_2.

Now $\alpha'\alpha''$ is a piecewise geodesic closed curve of length $\leq \ell_\alpha$. By Lemma 3.17.1 $\alpha'\alpha''$ is not homotopic to a point. Since $\alpha'\alpha''$ is not smooth, there is a closed geodesic curve β homotopic to $\alpha'\alpha''$ and

$$\ell_\beta < \ell_{\alpha'\alpha''} \leq \ell_\alpha.$$

This is not possible proving the lemma.

Let us consider in detail the case where α is of minimal length and α intersects $\sigma(\alpha)$ exactly at one point. Figure 3.11 shows a tubular distance

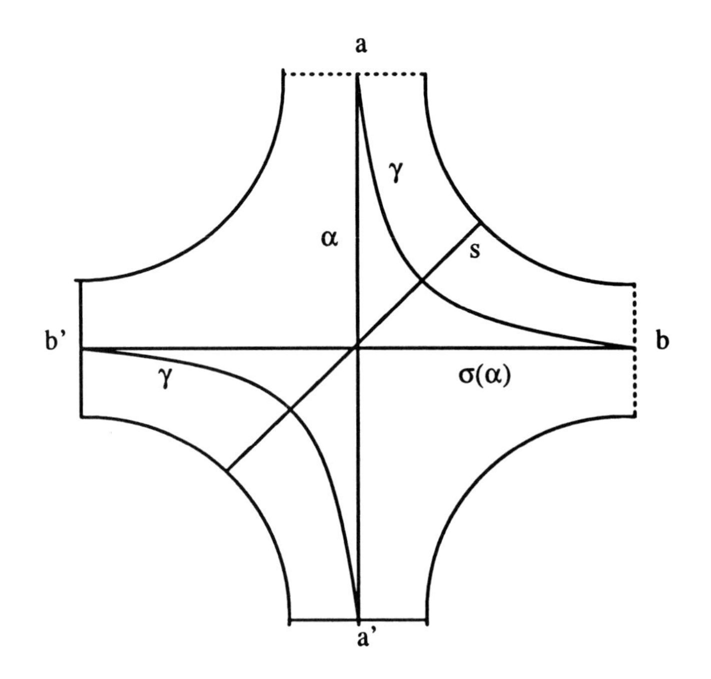

Figure 3.11: Tubular distance neighborhood of $\alpha \cup \sigma(\alpha)$.

neighborhood $T_\epsilon = N_\epsilon(\alpha \cup \sigma(\alpha))$ for some small positive distance ϵ. (The figure actually shows a fundamental domain of this neighborhood. Dotted line a has to be pasted to line a' in the obvious way so that x is identified with x', y with y' etc.) Similarly line b has to be pasted to line b'.

Since p is the only intersection point of α and $\sigma(\alpha)$ we must have $\sigma(p) = p$. Components of the fixed-point set of an antiholomorphic involution of a compact Riemann surface are always simple closed geodesic curves. We conclude therefore that the set of fixed-points of σ contains a geodesic arc s through the point p. Therefore the restriction $\sigma|_{T_\epsilon}$ of σ to T_ϵ is a reflection in the arc s.

Observe that the boundary of T_ϵ is a simple closed curve freely homotopic to $\alpha(\sigma(\alpha))^{-1}\alpha^{-1}\sigma(\alpha)$. Curve γ in the figure is a simple closed curve in the free homotopy class of $\alpha\sigma(\alpha)$. Observe that $\alpha\sigma(\alpha)$ is freely homotopic to $\sigma(\alpha\sigma(\alpha))$. (These two curves differ only by parametrization.)

By Epstein's theorem on isotopies, [28], see also [22], there exists an isotopy of X onto X which deforms $N_\epsilon(\alpha \cup \sigma(\alpha))$ into an embedded one-holed torus T with boundary geodesic γ_2 freely homotopic to $\alpha(\sigma(\alpha))^{-1}\alpha^{-1}\sigma(\alpha)$. The closed geodesic curve γ_1 which is freely homotopic to $\alpha\sigma(\alpha)$ is also simple and is contained in T.

We have $\sigma(T) = T$, $\sigma(\gamma_1) = \gamma_1$ and $\sigma(\gamma_2) = \gamma_2$. By inequality (3.38) we get

$$\ell_{\gamma_1} \leq 4\,\mathrm{arcosh}\,(2g-1) < 4\log(4g-2). \tag{3.39}$$

We remark that

$$\ell_{\gamma_2} \leq 8\log(4g-2). \tag{3.40}$$

although we do not need this in the following.

In conclusion we observe that the above considerations yield the following lemma.

Lemma 3.17.4 *Assume that X is a compact Riemann surface of genus g, $g > 1$, and $\sigma : X \to X$ an antiholomorphic involution. Let (α_1, α_2) be a σ-pair of minimal length ℓ_σ. If $\alpha_1 \cap \alpha_2 = \emptyset$, then this minimal length satisfies*

$$\ell_\sigma \leq 2\,\mathrm{arcosh}\,(2g-1) < 2\log(4g-2).$$

If $\alpha_1 = \alpha_2$, then

$$\ell_\sigma \leq 4\,\mathrm{arcosh}\,(2g-1) < 4\log(4g-2).$$

Observe that Lemma 3.17.4 implies the following result. Let (α_1, α_2) be a σ-pair with minimal length on X. $W_0 = X\backslash(\alpha_1 \cup \alpha_2)$ is a Riemann surface with boundary components and it need not be connected. The boundary of W_0 consists of geodesic curves and its length satisfies

$$\ell_{\partial W_0} \leq 8\,\mathrm{arcosh}\,(2g-1) < 8\log(4g-2). \tag{3.41}$$

Inequality (3.41) is an immediate consequence of (3.38) and (3.39)

3.18 Selection of additional simple closed curves on a hyperbolic Riemann surface with a symmetry

Above we have shown that it is always possible to select a geodesic σ–pair (α_1, α_2) on the symmetric Riemann surface X in such a way that cutting X open along this pair of simple closed curves yields a Riemann surface W_0 whose boundary ∂W_0 satisfies

$$\ell_{\partial W_0} \leq 8 \operatorname{arcosh}(2g-1). \tag{3.42}$$

Observe that for $g > 2$ (3.42) implies that

$$\ell_{\partial W_0} < 7g. \tag{3.43}$$

For $g = 2$ we have $\ell_{\partial W_0} < 7g + 0.102$.

We start with this Riemann surface W_0. In this section we show how to select new short essential σ–pairs inductively in such a way that they always cut out either a pair of pants or a four holed sphere.

At step k we consider the Riemann surface W_k (which need not be connected). From W_k we obtain a new Riemann surface W_{k+s} in the following manner.

On W_k we will select a σ–pair (β_1, β_2) that is not too long. The procedure for the selection of this σ–pair will be explained later together with an estimate for its length.

Let S_k be the union of all those components of $W_k \setminus (\beta_1 \cup \beta_2)$ that are pairs of pants or spheres with four holes. Let $s = -\chi(S_k)$, where $\chi(S_k)$ denotes the Euler characteristic of S_k. Put

$$W_{k+s} = W_k \setminus (S_k \cup \alpha_1 \cup \alpha_2).$$

Together with the σ–pair (α_1, α_2) we will find, on those components of S_k that are four holed spheres, additional curves β such that

- $\sigma(\beta) = \beta$.
- β satisfies the same length inequality than the total boundary of W_{k+2}.

It turns out that at each induction step we pass either from W_k to W_{k+1} or to W_{k+2} until we are left with a collection of pairs of pants. The numbering is made in such a way that at step k, when we are considering the Riemann surface W_k, we have already cut out k pairs of pants (with short boundary curves) from the original surface. Therefore this process necessarily stops with the Riemann surface W_{2g-2}.

If at some step we have produced a Riemann surface W_k for which the total length of the boundary, $\ell_{\partial W_k}$, is less than 2, then we use Corollary 3.14.9 and form the (maximal) collar at one of the boundary components of W_k. We replace W_k by the complement of this maximal collar. Corollary 3.14.9 implies then one of the boundary components of W_k has length at least 2 so that we have

$$\ell_{\partial W_k} > 2 \tag{3.44}$$

at each step of the induction.

Observe that this leads to Riemann surfaces whose boundaries are not geodesic curves. But this does not cause difficulties in our argument.

Let W_k be the Riemann surface of induction step k. Let $\alpha_1, \ldots, \alpha_n$ be the boundary components of W_k. They are simple closed curves on the Riemann surface W_k. Let $\ell_{\partial W_k}$ be the total length of these curves.

Recall that we use the notation ℓ_{α_j} to denote the length of the boundary component α_j. Then $\ell_{\partial W_k} = \sum_{j=1}^{n} \ell_{\alpha_j}$.

For $\epsilon > 0$ consider the set $N_\epsilon(\partial W_k)$. For small values of ϵ we have $N_\epsilon(\partial W_k) \approx \partial W_k \times [0, \epsilon)$.

Definition 3.18.1 *Let*

$$\epsilon_k = \sup\{\epsilon \mid N_\epsilon(\partial W_k) \approx \partial W_k \times [0, \epsilon)\}.$$

This number ϵ_k depends on the total length $\lambda_k := \ell_{\partial W_k}$ and plays a key role in our computations.

We have a number of different possible configurations of the set

$$N_{\epsilon_k}(\partial W_k).$$

Below is a list of all the possible configurations. We will consider the most complicated ones in detail. We have here divided the possible configurations into two main cases which then fall into subcases.

List of different configurations

1. It may happen that closures of two different components of $N_{\epsilon_k}(\partial W_k)$ meet. Assume that this is the case and call these components $N_{\epsilon_k}(\alpha_1)$ and $N_{\epsilon_k}(\alpha_2)$.

   This case falls further into subcases. They are:

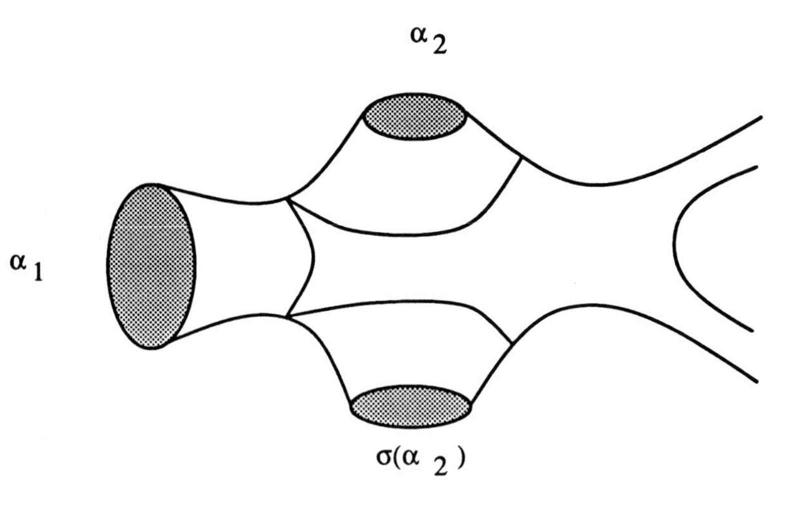

Figure 3.12: $\sigma(\alpha_1) = \alpha_1$ and $\sigma(\alpha_2) \neq \alpha_2$.

(a) $\sigma(\alpha_1) = \alpha_2$.
In this case also $\sigma(N_{\epsilon_k}(\alpha_1) \cup N_{\epsilon_k}(\alpha_2)) = N_{\epsilon_k}(\alpha_1) \cup N_{\epsilon_k}(\alpha_2)$. The new decomposing geodesic curve β will be the simple closed geodesic curve freely homotopic to the inner boundary of the set

$$N_{\epsilon_k}(\alpha_1) \cup N_{\epsilon_k}(\alpha_2).$$

If we cut the surface open along this curve β, we get one pair of pants and a new component W_{k+1}. The total length $\lambda_{k+1} := \ell_{\partial W_{k+1}}$ of the boundary of this new component satisfies

$$\lambda_{k+1} \leq \lambda_k + 4\epsilon_k. \tag{3.45}$$

(b) $\sigma(\alpha_1) = \alpha_1$ and $\sigma(\alpha_2) = \alpha_2$. In this case we may choose the new decomposing curve β in the same way as before. Equation (3.45) holds also in this case.

(c) $\sigma(\alpha_1) = \alpha_1$ and $\sigma(\alpha_2) \neq \alpha_2$. This is the worst case. Figure 3.12 illustrates this case. Let η be the geodesic arc of length $2\epsilon_k$ joining α_1 to α_2. By the definition of ϵ_k such an arc can be found. Let p be the end–point of η on α_1. Then p and $\sigma(p)$ divide α_1 into two arcs α_1' and α_1'', both having p and $\sigma(p)$ as end–points. Next observe[1] that since σ is an orientation reversing involution, $\sigma(\alpha_1') = \alpha_1'$ and $\sigma(\alpha_1'') = \alpha_1''$. The additional curve β is now

[1]This is a key point in our argument and follows from the fact thatσ is an orientation reversing self–mapping of α. Such a mapping has necessarily 2 fixed–points on α.

freely homotopic to the curve $\alpha_1'\eta\alpha_2\eta^{-1}\alpha_1''\sigma(\eta)\sigma(\alpha_2)^{-1}\sigma(\eta)^{-1}$. Here we assume that the orientations are chosen in a suitable way (cf. Fig. 3.12) so that β is a simple closed curve. Cutting the surface W_k open along this curve β yields one sphere with four holes S_k^4 and another component W_{k+2}. The total boundary length $\lambda_{k+2} := \ell_{\partial W_{k+2}}$ of this new component satisfies

$$\lambda_{k+2} \leq \lambda_k + 8\epsilon_k. \tag{3.46}$$

The four holed sphere S_k^4 can further be divided into two pairs of pants by the σ–invariant curve

$$\delta = \alpha_1'\eta\alpha_2\eta^{-1}\alpha_1'^{-1}\sigma(\eta)\sigma(\alpha_2)^{-1}\sigma(\eta)^{-1}.$$

Here we may change α_1' to α_1'' if necessary (i.e., if α_1'' is shorter than α_1'). The length of this new curve δ satisfies the inequality

$$\ell_\delta \leq \lambda_k + 8\epsilon_k.$$

(d) $\sigma(\alpha_1) \neq \alpha_1$ and $\sigma(\alpha_1) \neq \alpha_2$. Then also $\sigma(\alpha_2) \neq \alpha_1$ and $\sigma(\alpha_2) \neq \alpha_2$ because σ is an involution. Since

$$\overline{N_{\epsilon_k}(\alpha_1)} \cap \overline{N_{\epsilon_k}(\alpha_2)} \neq \emptyset$$

also

$$\overline{N_{\epsilon_k}(\sigma(\alpha_1))} \cap \overline{N_{\epsilon_k}(\sigma(\alpha_2))} \neq \emptyset.$$

Let β be the simple closed geodesic curve freely homotopic to the inner boundary component of the set $\overline{N_{\epsilon_k}(\alpha_1)} \cap \overline{N_{\epsilon_k}(\alpha_2)}$. We cut next the surface W_k open along the geodesic curves β and $\sigma(\beta)$. (Observe that it may very well happen that $\sigma(\beta) = \beta$.) We get *two* pairs of pants and a new surface W_{k+2}. The total length $\lambda_{k+2} := \ell_{\partial W_{k+2}}$ of the boundary of this new surface satisfies now the inequality

$$\lambda_{k+2} \leq \lambda_k + 8\epsilon_k. \tag{3.47}$$

2. In the second main case some component $\overline{N_{\epsilon_k}(\alpha_j)}$ touches itself. This means that $N_{\epsilon_k}(\alpha_j) \approx \alpha_j \times [0, \epsilon_k)$ but $\overline{N_{\epsilon_k}(\alpha_j)} \not\approx \alpha_j \times [0, \epsilon_k]$. Let γ be the inner boundary component of $N_{\epsilon_k}(\alpha_j)$. We have the following subcases:

   (a) $\sigma(\alpha_j) = \alpha_j$ and the inner boundary component γ is curve with a number of double points $p_1, \ldots, p_n$ none of which is fixed under σ.

   In this case γ can be decomposed into three parts. They are:

i. a simple closed curve γ_σ such that $\sigma(\gamma_\sigma) = \gamma_\sigma$,

ii. connected union γ_1 of simple closed subcurves of γ such that $\sigma(\gamma_1) \cap \gamma_1 = \emptyset$, and

iii. $\gamma_2 = \sigma(\gamma_1)$.

In this case, let β_1 be the simple closed geodesic curve freely homotopic to γ_1, $\beta_2 = \sigma(\beta_1)$ and β_3 the simple closed geodesic curve freely homotopic to γ_σ. Cut the surface W_k open along these curves. In this way one obtains a sphere with four holes S^4 and a new surface W_{k+2}. The total length λ_{k+2} of the boundary of W_{k+2} satisfies now the inequality (3.47).

The orientation reversing involution σ maps the curves α_j and γ_σ onto themselves. It follows that σ has, on both curves α_j and γ_σ two fixed–points and the fixed–point set of σ contains two (geodesic) arcs δ_1 and δ_2 which both connect the curves α_j and γ_σ. Let $A = \alpha_j \cup \delta_1 \cup \gamma_\sigma$. Then the set A is invariant under the involution σ. So is also any ϵ neighborhood $N_\epsilon(A)$ of A. For small values of ϵ, one boundary component of $N_\epsilon(A)$ is a σ–invariant simple closed curve contained in the four holed sphere S^4. This curve is homotopic to a simple closed geodesic curve β is S^4 which is invariant under σ, decomposes S^4 into two pairs of pants and whose length satisfies the inequality

$$\ell_\beta \leq \ell_{\alpha_j} + 4\epsilon_k \leq \lambda_k + 4\epsilon_k. \tag{3.48}$$

(b) $\sigma(\alpha_j) = \alpha_j$ and the inner boundary component γ is curve with a number of double points $p_1, \ldots, p_n$. At least one of them is fixed under σ. Assume that $\sigma(p_1) = p_1$. Then p_1 divides the inner boundary component into two parts (which are both taken to be closed curves with double points) γ_1 and γ_2. We either have $\sigma(\gamma_1) = \gamma_1$ or $\sigma(\gamma_1) = \gamma_2$. In this case let β_j be the simple closed curve freely homotopic to the curve γ_j. We cut the surface W_k open along these curves β_1 and β_2. We get one pair of pants and a new surface W_{k+1}. The total length λ_{k+1} of the boundary of W_{k+1} satisfies now the inequality (3.45).

(c) $\sigma(\alpha_j) \neq \alpha_j$. In this case again the inner boundary of $N_{\epsilon_k}(\alpha_j)$ decomposes into two parts that are freely homotopic to simple closed geodesic curves β_1 and β_2. Cut the surface open along the curves β_1, β_2, $\sigma(\beta_1)$ and $\sigma(\beta_2$. (NB It may happen that e.g. $\sigma(\beta_1) = \beta_1$. This does not pose any problems.)

This cutting yields us two pairs of pants and a new surface W_{k+2}. The total length λ_{k+1} of the boundary of W_{k+1} satisfies now the inequality (3.47).

In conclusion we observe that we have the following result:

Lemma 3.18.1 *Let W_k be a compact Riemann surface with boundary components and let $\sigma : W_k \to W_k$ be a antiholomorphic involution. Assume that $\chi(W_k) < -1$. Then it is always possible*

- *either to choose an essential geodesic σ-pair (β_1, β_2) which satisfies the length inequality given below and is such that $W_k \setminus (\beta_1 \cup \beta_2)$ has at least one component which is either a pair of pants or a four hole sphere*
- *or to choose a σ-invariant set $(\beta_1, \beta_2, \beta_3)$ of simple closed geodesic curves on W_k which satisfies the length inequality given below and is such that $W_k \setminus (\beta_1 \cup \beta_2 \cup \beta_3)$ has at least one component which is either a pair of pants or a four hole sphere.*

Let S_k be the union of all the components of $W_k \setminus \cup_j \beta_j$ that are spheres with either four or three holes. Let $s = \chi(S_k)$. The lengths of the curves β_j satisfy

$$\ell_{\beta_j} < \ell_{\partial W_k} + 4s\epsilon_k$$

where ϵ_k is the number given by Definition 3.18.1.

If one of the components of S_k is a four hole sphere S_k^4, then it can be decomposed into two pairs of pants by a σ-invariant simple closed geodesic curve β such that

$$\ell_\beta \leq \ell_{\partial W_k} + 8\epsilon_k$$

The length of each boundary curve γ of a pair of pants obtained at step k satisfies

$$\ell_\gamma < \ell_{\partial W_k} + 4s\epsilon_k.$$

Let $W_{k+s} = W_k \setminus S_k$. The length of the total boundary of W_{k+s} satisfies

$$\ell_{\partial W_{k+s}} < \ell_{\partial W_k} + 4s\epsilon_k.$$

3.19 Numerical estimate

A numerical estimate for the lengths of the decomposing curves can be obtained now by estimating the constants ϵ_k of Lemma 3.18.1.

Call the decomposing curves that we have obtained, by the above inductive process, $(\alpha_1, \alpha_2, \ldots, \alpha_{3g-3})$. Assume that the numbering corresponds also to the order of forming the curves.

Let W_0 be the first Riemann surface with boundary components which we obtained by cutting X open along a σ-pair with minimal length. By Lemma 3.17.4 we know that $\ell_{\partial W_0} < 8 \operatorname{arcosh}(2g-1)$.

Consider a Riemann surface W_k at some induction step k. If $\ell_{\partial W_k} < 2$, then we delete the maximal disjoint collars at boundary components and replace the original surface W_k with this new Riemann surface. We may, therefore, suppose that $\ell_{\partial W_k} > 2$. Hence we have

$$4\pi(g-1) > \text{Area of } N_{\epsilon_k}(\partial W_k) \geq \ell_{\partial W_k} \sinh \epsilon_k \geq 2 \sinh \epsilon_k.$$

We conclude in this case that

$$\epsilon_k \leq \operatorname{arsinh}(2\pi(g-1)) := T_1. \tag{3.49}$$

We can improve our estimate concerning ϵ_k if we assume that $\ell_{\partial W_k} \geq 7g$. Then, by repeating the above reasoning concerning the area of $N_{\epsilon_k}(\partial W_k)$, we obtain

$$\epsilon_k \leq \operatorname{arsinh}\left(\frac{4\pi(g-1)}{7g}\right) := T_2. \tag{3.50}$$

By Lemma 3.18.1 we have the following upper bound for the lengths of all the boundary components α_{k_j} obtained at step k:

$$\ell_{\alpha_{k_j}} \leq \ell_{\partial W_k} + 4\epsilon_k \tag{3.51}$$

or

$$\ell_{\alpha_{k_j}} \leq \ell_{\partial W_k} + 8\epsilon_k \tag{3.52}$$

according to as whether the induction step goes from W_k to W_{k+1} or to W_{k+2}. By the same lemma we also have

$$\ell_{\partial W_{k+1}} \leq \ell_{\partial W_k} + 4\epsilon_k \tag{3.53}$$

or

$$\ell_{\partial W_{k+2}} \leq \ell_{\partial W_k} + 8\epsilon_k. \tag{3.54}$$

Since $\ell_{\partial W_0} < 7g + 1$, we get, expecting the worst,

$$\ell_{\partial W_k} \leq 7g + 1 + 8T_1 + 4(k-2)T_2 \tag{3.55}$$

$$\ell_{\alpha_{k_j}} \leq 1 + 7g + 8T_1 + 4kT_2 \tag{3.56}$$

with an obvious modification for $k = 0, 1, 2$. The last possible step at which we may find new curves is $k = 2g - 3$. This yields the upper bound

$$\ell_{\alpha_{k_j}} \leq 1 + 7g + 8T_1 + 4(2g-3)T_2 \leq 21g.$$

This proves the following theorem:

Theorem 3.19.1 *Let X be a compact Riemann surface of genus g, $g > 1$. Assume that $\sigma : X \to X$ is an antiholomorphic involution. Then X has a σ–invariant set of $3g - 3$ disjoint simple closed geodesic curves such that each curve is of length $< 21g$.*

One should observe here that the above upper bound $21g$ is not sharp. A closer examination of the above arguments allows the reader to find a lower upper bound for the lengths of the $3g - 3$ curves.

We conclude by observing that the symmetry σ played no role at all in Theorem 3.19.1. It only made the proof more complicated. Therefore we have also the following result:

Theorem 3.19.2 *Let X be a compact Riemann surface of genus g, $g > 1$. X has a set of $3g - 3$ disjoint simple closed geodesic curves such that each curve is of length $< 21g$.*

3.20 Groups of Möbius transformations and matrix groups

Let X be a Riemann surface of genus g, $g > 1$. By the Uniformization we may express X as the quotient U/G for some Fuchsian group G. Such a Fuchsian group is a subgroup of $\mathrm{PSL}_2(\mathbf{R})$, the group of holomorphic automorphisms of the upper half-plane.

Elements of $\mathrm{PSL}_2(\mathbf{R})$ are of the form

$$g(z) = \frac{az+b}{cz+d}, \quad \text{with } ad - bc = +1.$$

Therefore we can associate a matrix

$$\tilde{g} = \begin{pmatrix} a & b \\ c & d \end{pmatrix} \tag{3.57}$$

in $\mathrm{SL}_2(\mathbf{R})$ to each Möbius transformation $g : U \to U$. This is, of course, standard but not quite unique. Both matrices $\tilde{g}$ and $-\tilde{g}$ correspond to the same Möbius transformation.

Let g and h be two Möbius transformations in $\mathrm{PSL}_2(\mathbf{R})$. Let $\tilde{g}$ and $\tilde{h}$ be the associated matrices. Then $\tilde{c} = \tilde{h}\tilde{g}^{-1}\tilde{h}^{-1}\tilde{g}$ is a matrix corresponding to the commutator $c = h \circ g^{-1} \circ h^{-1} \circ g$ of h and g. Observe that even though the matrices $\tilde{h}$ and $\tilde{g}$ corresponding to the mappings h and g are determined only up to sign, the matrix $\tilde{c}$ depends neither on the choice of the sign of $\tilde{h}$ nor on that of $\tilde{g}$. So we may speak of *the matrix* of the commutator c.

In this and the subsequent sections we study traces of matrices of the commutators c of hyperbolic Möbius transformations h and g whose axes intersect. In the next section we show that for hyperbolic commutators c the trace of the corresponding matrix $\tilde{c}$ is always negative.

Let $g_1, \ldots, g_{2p}$ be the standard set of generators for the Fuchsian group G. Let $c_j = [g_{2j-1}, g_{2j}] = g_{2j} \circ g_{2j-1}^{-1} \circ g_{2j}^{-1} \circ g_{2j-1}$ be the commutator of g_{2j-1} and g_{2j} and let $\tilde{c}_j$ be its matrix. It turns out that the traces of the matrices

$$\tilde{c}_j \tilde{c}_{j-1} \cdots \tilde{c}_1$$

are all negative for $j < p$ and equal to $+2$ for $j = p$.

A corollary of this result is that a Fuchsian subgroup of $\mathrm{PSL}_2(\mathbf{R})$ can always be lifted to a subgroup of $\mathrm{SL}_2(\mathbf{R})$. This is not obvious since the exact sequence

$$1 \to \{\pm I\} \to \mathrm{SL}_2(\mathbf{R}) \xrightarrow{\pi} \mathrm{PSL}_2(\mathbf{R}) \to 1 \tag{3.58}$$

does not split. The fact that Fuchsian groups lift has actually been proven several times by many methods and authors. An interesting survey of the history of the problem together with an elegant proof for a generalization of this result can be found in Irwin Kra's paper [53]. Here we follow the arguments presented in [82].

3.21 Traces of commutators

We use the classification of pairs of Möbius transformations that was introduced in Chapter 1, Section 1.4. (Recall the notation from page 24, we use the same notation here.)

Let g and h be Möbius transformations fixing the upper half-plane U, and let $\tilde{g}$ and $\tilde{h}$ be some matrices in $\mathrm{SL}_2(\mathbf{R})$ corresponding to g and h. Then the matrix product $\tilde{g}\tilde{h}$ is a representation of $g \circ h$. Denote the trace of a matrix A by $\chi(A)$ and observe that $\chi(\tilde{g}\tilde{h}) = \chi\big((-\tilde{g})(-\tilde{h})\big)$.

Lemma 3.21.1 $\chi(\tilde{g}) = \chi(\tilde{h}\tilde{g}\tilde{h}^{-1})$.

Proof. Let $\tilde{g} = \begin{pmatrix} a & b \\ c & d \end{pmatrix}$. A straightforward multiplication of matrices shows that

$$\chi(\tilde{h}\tilde{g}\tilde{h}^{-1}) = (\det\tilde{h})\chi(\tilde{g}) = \chi(\tilde{g})$$

proving the lemma.

Suppose that g and h are hyperbolic transformations without common fixed points.

Lemma 3.21.2 *If* $(g, h) \in \operatorname{Int} \mathcal{P}$ *and* $\chi(\tilde{g})\chi(\tilde{h}) > 0$, *then* $\chi(\tilde{g}\tilde{h}) < 0$.

Proof. By Lemma 3.21.1, we may suppose that $a(h) = 0$, $r(h) = 1$ and $a(g) = \infty$. Then

$$t = r(g) < t_2 = \frac{-2 - f(k_1/k_2)}{f(k_1k_2) - f(k_1/k_2)}$$

and

$$\begin{aligned} g(z) &= k_1 z - t(k_1 - 1), \; k_1 = k(g) > 1, \\ h(z) &= \frac{z}{(1-k_2)z + k_2}, \; k_2 = k(h) > 1. \end{aligned}$$

Hence

$$\tilde{g} = \begin{pmatrix} \sqrt{k_1} & -t(\sqrt{k_1} - \frac{1}{\sqrt{k_1}}) \\ 0 & \frac{1}{\sqrt{k_1}} \end{pmatrix}, \; \chi(\tilde{g}) > 0,$$

$$\tilde{h} = \begin{pmatrix} \frac{1}{\sqrt{k_2}} & 0 \\ -(\sqrt{k_2} - \frac{1}{\sqrt{k_2}}) & \sqrt{k_2} \end{pmatrix}, \; \chi(\tilde{h}) > 0.$$

It follows that

$$\begin{aligned} \chi(\tilde{g}\tilde{h}) &= \frac{\sqrt{k_1}}{\sqrt{k_2}} + t(\sqrt{k_1} - \frac{1}{\sqrt{k_1}})(\sqrt{k_2} - \frac{1}{\sqrt{k_2}}) + \frac{\sqrt{k_2}}{\sqrt{k_1}} \\ &= tf(k_1k_2) + (1-t)f\left(\frac{k_1}{k_2}\right) < -2. \end{aligned}$$

Suppose now that g and h have intersecting axes.

Lemma 3.21.3 *The commutator*

$$c = [g, h] = h \circ g^{-1} \circ h^{-1} \circ g$$

is hyperbolic if and only if $(h, g^{-1} \circ h^{-1} \circ g) \in \operatorname{Int} \mathcal{P}$.

Proof. Denote $h' = g^{-1} \circ h^{-1} \circ g$. To consider the class[2] of the pair (h, h'), let $t = (r(h), r(h'), a(h'), a(h))$ and $k = k(h) = k(h')$. Recall that here $r(h)$, $r(h')$ are the repelling fixed points and $a(h)$, $a(h')$ are the attracting fixed points while k is the common multiplier of h and h'.

Then

$$\begin{aligned} t_1 &= \frac{2 - f(k/k)}{f(k^2) - f(k/k)} &= 0, \\ t_2 &= \frac{-2 - f(k/k)}{f(k^2) - f(k/k)} &= \frac{-4}{f(k^2) - 2} < 0. \end{aligned}$$

[2]For the definition of the class of a pair of Möbius transformations see page 1.4.

By Lemma 3.21.1, we may suppose that $a(h) = \infty$, $a(h') = 0$, $r(h') = 1$ and $r(h) = t$. Since

$$\begin{aligned} g^{-1}(r(h)) &= a(h') &= 0 \\ g^{-1}(a(h)) &= r(h') &= 1, \end{aligned}$$

we have $g(0) = t$ and $g(1) = \infty$. On the other hand, the axis of g intersects with the axis of h and that of h'. From $g(1) = \infty$ it then follows that

$$a(g) < t < 0 < r(g) < 1.$$

The commutator $c = h \circ h'$ is hyperbolic if and only if $(h, h') \in \operatorname{Int}\mathcal{P} \cup \operatorname{Int}\mathcal{H}$, i.e., if and only if $t < t_2$ or $t_1 < t$. Since $t_1 = 0$ and $t < 0$, the assertion follows.

Lemma 3.21.4 *If $c = [g, h]$ is hyperbolic and $\tilde{c} = \tilde{h}\tilde{g}^{-1}\tilde{h}^{-1}\tilde{g}$, then $\chi(\tilde{c}) < 0$.*

Proof. Since $\chi(\tilde{h}) = \chi(\tilde{h}^{-1})$, we have $\chi(\tilde{h})\chi(\tilde{g}^{-1}\tilde{h}^{-1}\tilde{g}) > 0$ by Lemma 3.21.1. By Lemma 3.21.3, $(h, g^{-1} \circ h^{-1} \circ g) \in \operatorname{Int}\mathcal{P}$. Hence the pair $(h, g^{-1} \circ h^{-1} \circ g)$ fulfills the assumptions of Lemma 3.21.2 for any choice of the representations $\tilde{g}$ and $\tilde{h}$ of g and h, and the assertion follows.

Let G be a Fuchsian group acting in U. Suppose that U/G is a compact Riemann surface of genus p. Suppose that $g \in G$ and $h \in G$ correspond to simple closed geodesics α and β of U/G, respectively. If α and β do not intersect, then either $(g, h) \in \operatorname{Int}\mathcal{P}$ or $(g, h^{-1}) \in \operatorname{Int}\mathcal{P}$ depending on the cyclic order of the fixed points of g and h. (This follows from the considerations on page 25, see also [80].) Let $g_1, g_2, \ldots, g_{2p}$ be a canonical set of generators of G. Let $\tilde{g}_1, \tilde{g}_2, \ldots, \tilde{g}_{2p}$ be representations of the generating transformations in $\mathrm{SL}_2(\mathbf{R})$. Then $\tilde{c}_j = [\tilde{g}_{2j-1}, \tilde{g}_{2j}]$ is a representation of c_j. Since U/G is compact, G contains, besides the identity, only hyperbolic elements (Theorem 3.12.1). Hence by Lemma 3.21.4, $\chi(\tilde{c}_j) < 0$. For any $j = 1, \ldots, p-2$, the transformations

$$c_j \circ c_{j-1} \circ \cdots \circ c_1, \quad c_{j+1}, \ldots, c_p$$

correspond to simple closed geodesics on U/G. Moreover, all these geodesics are pairwise disjoint and $(c_{j+1}, c_j \circ \cdots \circ c_1) \in \operatorname{Int}\mathcal{P}$.

Theorem 3.21.5 $\chi(\tilde{c}_j\tilde{c}_{j-1}\cdots\tilde{c}_1) < 0$ *for* $j = 1, \ldots, p-1$.

Proof. For $j = 1$ the assertion holds by Lemma 3.21.4. Suppose that $\chi(\tilde{c}_{j-1}\cdots\tilde{c}_1) < 0$. Since $\chi(\tilde{c}_j) < 0$ and $(c_j, c_{j-1} \circ \cdots \circ c_1) \in \operatorname{Int}\mathcal{P}$, we have, by Lemma 3.21.2, $\chi(\tilde{c}_j\tilde{c}_{j-1}\cdots\tilde{c}_1) < 0$.

If we choose $j = p-1$, then we have

- $\chi(\tilde{c}_{p-1}\tilde{c}_{p-2}\cdots\tilde{c}_1) < 0$,
- $\chi(\tilde{c}_p) < 0$,
- $(c_{p-1}, c_{p-2} \circ \cdots \circ c_1) \in \operatorname{Int} \mathcal{P}$,
- $c_p = (c_{p-1} \circ \cdots \circ c_1)^{-1}$.

Theorem 3.21.6 $\chi(\tilde{c}_p\tilde{c}_{p-1}\cdots\tilde{c}_1) = 2$.

Proof. Since $c_p \circ c_{p-1} \circ \cdots \circ c_1 = id$, we have $\tilde{c}_p\tilde{c}_{p-1}\cdots\tilde{c}_1 = \pm$Identity.

By Lemmata 3.21.1 and 3.21.4, we may suppose that

$$\tilde{c}_p = \begin{pmatrix} -a & 0 \\ 0 & -a^{-1} \end{pmatrix}, \quad a > 1.$$

Then, by Theorem 3.21.5, $\tilde{c}_{p-1}\tilde{c}_{p-2}\cdots\tilde{c}_1 = \tilde{c}_p^{-1}$. Hence we have in fact

$$\tilde{c}_p\tilde{c}_{p-1}\cdots\tilde{c}_1 = \text{Identity}. \tag{3.59}$$

3.22 Liftings of Fuchsian groups

Let $G \subset \mathrm{PSL}_2(\mathbf{R})$ be a group of Möbius transformations. Consider the exact sequence (3.58). We say that a subgroup $G \subset \mathrm{PSL}_2(\mathbf{R})$ *can be lifted to* $\mathrm{SL}_2(\mathbf{R})$ if there exists a subgroup $\Gamma \subset \mathrm{SL}_2(\mathbf{R})$ such that $\pi : \Gamma \to G$ is an isomorphism.

Let G be the group generated by the elliptic Möbius transformation $g(z) = -1/z$ of order two. It is immediate that this group cannot be lifted to $\mathrm{SL}_2(\mathbf{R})$ because the matrix of g,

$$\begin{pmatrix} 0 & -1 \\ 1 & 0 \end{pmatrix}$$

is of order 4 while g is of order 2. In [53] Irwin Kra proves the more general result stating that a subgroup $G \subset \mathrm{PSL}_2(\mathbf{R})$ can be lifted to $\mathrm{SL}_2(\mathbf{R})$ if and only if G does not have any elements of order 2 ([53, Theorem p. 181]).

Kra's proof is based on the existence of square roots of the canonical bundle of a compact Riemann surface. He treats the general case with Maskit's combination theorem.

Let $g_1, g_2, \ldots, g_{2p}$ be standard generators for a Fuchsian group G as before. Then all generators g_j and commutators c_j are hyperbolic Möbius transformations. Let $\tilde{g}_j$ be *any* matrix corresponding to the Möbius transformation g_j. Let $\tilde{c}_j = [\tilde{g}_{2j-1}, \tilde{g}_{2j}]$, $j = 1, 2, \ldots, p$.

Consider the group $\Gamma = \langle \tilde{g}_1, \ldots, \tilde{g}_p \rangle$. By Theorem 3.21.6 the matrices $\tilde{c}_j$ satisfy the relation 3.59 for any choice of the matrices $\tilde{g}_j$. It is also obvious that there cannot be any other relations among the generators $\tilde{g}_j$ of Γ because any such relation would imply a new relation among the generators g_j of G. We conclude that the groups Γ and G are isomorphic and that the restriction of the projection $\pi : \mathrm{SL}_2(\mathbf{R}) \to \mathrm{PSL}_2(\mathbf{R})$ to Γ is an isomorphism. Therefore we have:

Theorem 3.22.1 *Genus p, $p > 1$, Fuchsian subgroups of* $\mathrm{PSL}_2(\mathbf{R})$ *have* 2^{2p} *different liftings to* $\mathrm{SL}_2(\mathbf{R})$.

Chapter 4

Moduli problems and Teichmüller spaces

4.1 Introduction to Chapter 4

The famous *Riemann moduli problem* is the following: *Characterize the set of isomorphism classes of Riemann surfaces of a given topological type.*

Riemann solved this problem for simply connected Riemann surfaces. The Riemann mapping theorem states that there are only three different types of simply connected Riemann surfaces; namely the Riemann sphere, the finite complex plane and the unit disk.

Considerations presented for the definition of quasiconformal mappings in Chapter 2 imply that a doubly connected Riemann surface X is always a ring domain of type $A(r_X, R_X) = \{z \in \mathbf{C} \mid r_X < |z| < R_X\}$ for some R_X, r_X, $\infty \geq R_X > r_X \geq 0$. The quotient R_X/r_X determines the isomorphism class of a doubly connected Riemann surface X. Therefore the *moduli space of doubly connected Riemann surfaces*, i.e., the space of isomorphism classes of doubly connected Riemann surfaces can be viewed as the infinite half–line $\{r \mid r > 1\} \cup \{\infty\}$. This is actually not quite accurate: $\{z \in \mathbf{C} \mid 0 < |z| < 1\}$ is not isomorphic to $\{z \in \mathbf{C} \mid 0 < |z| < \infty\}$ but for both of them R_X/r_X should be defined ot be ∞. We do not consider this technical complication here.

The situation becomes more complicated as the topology of the Riemann surface in question gets richer. First such complications are encountered already with triply connected Riemann surfaces, i.e., with pairs of pants. Let now P^2 be a topological pair of pants. We consider P^2 as being an *oriented* closed Riemann surface with three boundary components which are simple closed curves. Let

$$\mathcal{M}(P^2) = \{X \mid X \text{ a complex structure on } P^2\}.$$

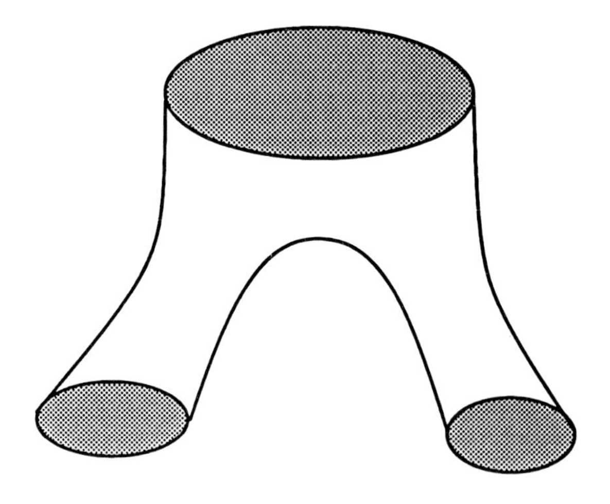

Figure 4.1: A pair of pants.

We assume that the complex structures $X \in \mathcal{M}(P^2)$ agree with the orientation of P^2 and make (P^2, X) a Riemann surface with three boundary components. Write $X = (P^2, X) = U/G$ for a Fuchsian group G.

The fundamental group $\pi_1(P^2, p)$ of P^2 at some base–point $p \in P^2$ is freely generated by the homotopy classes of two simple closed curves α and β going around two of the three holes of P^2. Provided that the orientations are suitably chosen, $\alpha\beta$ is a simple closed curve going around the third hole. Assume that this is the case.

Let $X = (P^2, X) = U/G$. Then G is a Fuchsian group freely generated by two Möbius transformations g and h. The group G is also isomorphic to $\pi_1(P^2, p)$ and we assume that the isomorphism $i : \pi_1(P^2, p) \to G$ is such that $i(\langle\alpha\rangle) = g$ and $i(\langle\beta\rangle) = h$.

By considerations related to Theorem 3.12.1 (see page 107) we conclude that the elements of $G \setminus \{Id\}$ are hyperbolic Möbius transformations (since we assumed that P^2 is a compact surface with boundary components).

Since α and β are simple closed curves, the hyperbolic transformations g and h are simple or primary elements of G (see page 33). By Theorem 1.5.6, either (g, h) or (g^{-1}, h) is in the class $\mathcal{P}$ depending on the orientation of α and β. Since $\alpha\beta$ is a simple closed curves, it follows that (g, h) satisfies (1.26) (on page 33) and that in fact (g, h) belongs to the class $\operatorname{Int}\mathcal{P}$.

Assume now that X' is another complex structure of P^2 which agrees with the orientation of P^2. Let $X' = (P^2, X') = U/G'$. Then G' is a Fuchsian group freely generated by two Möbius transformations g' and h' corresponding to $\langle\alpha\rangle \in \pi_1(P^2, p)$ and to $\langle\beta\rangle \in \pi_1(P^2, p)$. This means that the identity mapping $P^2 \to P^2$ induces a mapping $f : X \to X'$ that has a lifting $F : U \to U$ such that

$$F \circ g = g' \circ F \text{ and } F \circ h = h' \circ F. \tag{4.1}$$

Recall that by Lemma 1.4.1 of Chapter 1, the pair $(g, h) \in \mathcal{P}$ is determined, up to a conjugation by a Möbius transformation, by the parameters $k(g)$, $k(h)$ and $k(g \circ h)$. The Möbius transformations g and h cover homotopy classes of curves going around two of the boundary components of P^2 (cf. Definition 3.6.1). The Möbius transformation $g \circ h$ covers the homotopy class of a curve going around the third boundary component, and $\log k(g)$, $\log k(h)$ and $\log k(g \circ h)$ are the lengths of the geodesic curves homotopic to the boundary components of X (measured in the complete hyperbolic metric of X, not in the intrinsic metric). Here $k(g)$, $k(h)$ and $k(g \circ h)$ are the multipliers of the corresponding Möbius transformations.

In view of Lemma 1.4.1 and Lemma 3.6.7 we have now the following result:

Lemma 4.1.1 *The homotopy class of the identity mapping $(P^2, X) \to (P^2, X')$ contains a holomorphic homeomorphism if and only if the geodesic curves homotopic to the boundary components of P^2 have the same length on (P^2, X) and on (P^2, X').*

Lemma 4.1.1 leads to the following construction. Observe that the group $\text{Homeo}(P^2)$ of homeomorphic self-mappings of P^2 acts on $\mathcal{M}(P^2)$ in the following way. Let $X \in \mathcal{M}(P^2)$ and $f \in \text{Homeo}(P^2)$. If f is orientation preserving, then $f^*(X) \in \mathcal{M}(P^2)$ is defined as that complex structure of P^2 for which $f : (P^2, f^*(X)) \to (P^2, X)$ is a holomorphic homeomorphism. For orientation reversing mappings f, $f : (P^2, f^*(X)) \to (P^2, X)$ is required to be antiholomorphic.

The subgroup $\text{Homeo}_0(P^2)$ of $\text{Homeo}(P^2)$, which consists of mappings homotopic to the identity mapping, acts also on $\mathcal{M}(P^2)$.

Definition 4.1.1 *The* Teichmüller space *of a pair of pants P^2 is*

$$T(P^2) = \mathcal{M}(P^2)/\text{Homeo}_0(P^2).$$

Here the Teichmüller space $T(P^2)$ is defined only as a set. The definition is motivated by Lemma 4.1.1 which implies that the mapping

$$T(P^2) \to \mathbf{R}^3_+, \ [X] \mapsto (\ell_\alpha(X), \ell_\beta(X), \ell_\gamma(X)), \tag{4.2}$$

is injective. It is also onto by Lemma 1.4.1. Here α, β and γ denote the boundary curves of P^2, $\ell_\alpha(X)$, $\ell_\beta(X)$ and $\ell_\gamma(X)$ are the lengths of the geodesic curves on (P^2, X) homotopic to these boundary curves and $\mathbf{R}_+ = \{r \in \mathbf{R} \mid r > 0\}$. Therefore, the Teichmüller space of a pair of pants P^2 is simply the space $\mathbf{R}^3_+$ which is a rather simple 3 dimensional real manifold.

Definition 4.1.2 *The* moduli space *of P^2, $M(P^2)$, is the set of isomorphism classes of complex structures of P^2. In other words,*

$$M(P^2) = \mathcal{M}(P^2)/\mathrm{Homeo}_+(P^2),$$

where $\mathrm{Homeo}_+(P^2)$ *is the subset of* $\mathrm{Homeo}(P^2)$ *consisting of orientation preserving mappings.*

Topological arguments show that an orientation preserving mapping $f : P^2 \to P^2$ is homotopic to the identity if and only if f keeps the boundary components fixed (a sets, not necessarily point–wise). It is, on the other hand, clear that $\mathrm{Homeo}_+(P^2)$ contains mappings that permute the boundary components in all the 6 possible ways. We conclude, therefore, that

$$M(P^2) = (\mathbf{R}_+)^3/\text{perm. of coord.} \tag{4.3}$$

This simple example of pairs of pants serves to illustrate the differences between the *Teichmüller space* and the *moduli space* of a surface. The ideas that lead to the definition of the Teichmüller space, given by Lars V. Ahlfors on October 18, 1953 (cf. [4, § VIII, Page 53]), were originally due to Oswald Teichmüller ([94]). He called these spaces 'moduli spaces' and proposed that the original Riemann moduli space should be replaced by this new moduli space which is a covering of the Riemann moduli space. Today the moduli space originally defined by Teichmüller is known as the Teichmüller space and the Riemann moduli space is usually referred to as the moduli space.

There are deep reasons why the Teichmüller spaces of compact surfaces are, in some sense, easier to study than the corresponding moduli spaces. One of these is the famous *Extremal Mapping Theorem* of Teichmüller ([93], cf. 4.2.3). That result states that, within each homotopy class of homeomorphisms between compact Riemann surfaces of genus $g > 1$ there is always a *unique extremal* quasiconformal mapping. (Extremal in the sense that its maximal dilatation is the smallest in its homotopy class.) This result will be explained (but not shown) in more detail in the proceeding sections. It opened new avenues to attack the moduli problem.

In this monograph we will only discuss some of the consequence of the Teichmüller extremal mapping theorem. We will not prove it here. The original proof of Teichmüller ([93]) is very difficult to read. First actual proof for this theorem was given by Ahlfors in [4]. A good presentation of this theorem can be found for instance in the monograph of Olli Lehto[60].

4.2 Quasiconformal mappings of Riemann surfaces

In this section we extend the definitions related to quasiconformal mappings to mappings between Riemann and Klein surfaces. We will review basic results concerning these mappings. This serves to describe the usual constructions for the topology and complex structure of Teichmüller spaces. Our main interest is to study Teichmüller spaces using the methods of Chapter 1. This gives us information about the topology and real analytic geometry of Teichmüller spaces. The deliberations of this and the subsequent chapters serve only to explain the connections to the more classical approach (cf. e.g. [60]) that uses quasiconformal mappings.

Definition 4.2.1 *A homeomorphism $f : X \to X'$ between Klein surfaces is* K–quasiconformal *if it is locally K-quasiconformal in the following sense: for each point $P \in X$ there is a connected open neighborhood U of P, a dianalytic local variable $z : U \to z(U) \subset \mathbf{C}$ and a dianalytic local variable $w : V \to w(V)$ of a neighborhood V of $f(P)$ such that the mapping $w \circ f \circ z^{-1}$ is a K-quasiconformal homeomorphism of a neighborhood of $z(P)$.*

Since Klein surfaces need not be orientable, we cannot speak of orientation preserving mappings between them. This causes minor technical difficulties and makes some considerations and formulae cumbersome.

Computing locally we may form the Beltrami differential (cf. Definition 2.4.4) of a quasiconformal mapping $f : X \to X'$ between Klein surfaces: If $\{(U_i, z_i) |\, i \in I\}$ is a dianalytic atlas of X (whose charts U_i are connected), then for each index $i \in I$ there is a dianalytic chart (V_{j_i}, w_{j_i}) of X' such that $f(U_i) \subset V_{j_i}$ and $w_{j_i} \circ f \circ z_i^{-1}$ is quasiconformal in $z_i(U_i)$.

Let τ_i be the complex dilatation of this quasiconformal mapping. Consider the family of functions $\mu_i = \tau_i \circ z_i^{-1}$ associated to the different dianalytic charts of X. The function μ_i is a complex valued L^∞–function defined on U_i and $\|\mu_i\|_\infty \le \frac{K-1}{K+1} < 1$.

To see how the different functions μ_i are related to each other, consider two intersecting dianalytic charts (U_i, z_i) and (U_j, z_j) of X. Define the function $T_{ij} : U_i \cap U_j \to \mathbf{C}$ setting

$$T_{ij} = (\partial(z_i \circ z_j^{-1}) + \overline{\partial}(z_i \circ z_j^{-1})) \circ z_j.$$

Here we have used the usual notations

$$\overline{\partial} = \frac{\partial}{\partial \overline{z}} = \frac{1}{2}\left(\frac{\partial}{\partial x} + i\frac{\partial}{\partial y}\right)$$

and

$$\partial = \frac{\partial}{\partial z} = \frac{1}{2}\left(\frac{\partial}{\partial x} - i\frac{\partial}{\partial y}\right).$$

The following transformation formula is a straightforward computation: If $z_i \circ z_j^{-1}$ is holomorphic at $z_j(P)$, $P \in U_i \cap U_j$, then

$$\mu_j(P) = \mu_i(P)\frac{\overline{T_{ij}(P)}}{T_{ij}(P)}. \tag{4.4}$$

If $z_i \circ z_j^{-1}$ is antiholomorphic at $z_j(P)$, then

$$\overline{\mu_j(P)} = \mu_i(P)\frac{\overline{T_{ij}(P)}}{T_{ij}(P)}. \tag{4.5}$$

Definition 4.2.2 *We say that a collection $\mu = \{\mu_i | \, i \in I\}$ of measurable functions $\mu_i : U_i \to \mathbf{C}$ associated to dianalytic charts (U_i, z_i) of X is a (-1,1)-*differential of the Klein surface X *if the functions μ_i satisfy the transformation rules (4.4) and (4.5) (on page 142). If, in addition,*

$$\sup_{i \in I} ||\mu_i||_\infty < 1$$

then μ is a Beltrami differential of the Klein surface X. *Let us use the notation $D^{(-1,1)}(X)$ for the space of (-1,1)-differentials of X and the notation $Bel(X)$ for the space of the Beltrami differentials of X.*

For a Riemann surface X, (i.e., for an orientable Klein surface X) $D^{(-1,1)}(X)$ is a complex Banach-space and $Bel(X)$ is its open unit ball. If X is not orientable, then, since the transformation rule (4.5) has to hold, elements of $Bel(X)$ cannot be multiplied by complex numbers. Hence, for a non-orientable Klein surface X, $Bel(X)$ is the open unit ball of a real Banach space.

By the above remarks, the complex dilatation of a quasiconformal mapping of a Klein surface X is a Beltrami differential of X. The following theorem is an immediate application of the existence and the uniqueness theorems of plane quasiconformal mappings:

Theorem 4.2.1 *Let μ be a Beltrami differential of a Klein surface $X = (\Sigma, X)$ then there exists a dianalytic structure X_μ of the topological surface Σ such that the identity mapping $(\Sigma, X) \to (\Sigma, X_\mu)$ is quasiconformal with the complex dilatation μ.*

If $f_i : X \to f_i(X)$, $i = 1, 2$, are two μ-quasiconformal mappings of X, then there exists an isomorphism $g : f_1(X) \to f_2(X)$ such that $f_2 = g \circ f_1$.

Proof. Let $\mathcal{U} = \{(U_i, z_i) \mid i \in I\}$ be a dianalytic atlas of (Σ, X). Let $\mu_i : U_i \to \mathbf{C}$ be the measurable function associated to the chart (U_i, z_i) by the Beltrami differential μ.

By Theorem 2.4.6 there exists $\mu \circ z_i^{-1}$–quasiconformal mappings $f_\mu^i : z_i(U_i) \to \mathbf{C}$. Choose one for each index $i \in I$.

Then by the transformation rule (2.8) $\mathcal{U}_\mu = \{(U_i, f_\mu^i \circ z_i \mid i \in I\}$ is dianalytic atlas of Σ. Let X_μ be the dianalytic structure defined by this atlas. By the definition it is now clear that the identity mapping $(\Sigma, X) \to (\Sigma, X_\mu)$ is μ–quasiconformal proving the first statement.

The second statement follows directly from the transformation rule (2.8).

Let $X = U/G$ where G is a reflection group. A (–1,1)–differential μ of X lifts to a function $\mu : U \to \mathbf{C}$. The transformation rules (4.4) and (4.5) are equivalent with the following formulae:

$$\mu = (\mu \circ g)\frac{\overline{\partial g}}{\partial g} \tag{4.6}$$

for orientation preserving Möbius–transformations $g \in G$ and

$$\overline{\mu} = (\mu \circ \sigma)\frac{\overline{\overline{\partial}\sigma}}{\overline{\partial}\sigma} \tag{4.7}$$

for glide–reflections $\sigma \in G$.

Definition 4.2.3 *A measurable function μ satisfying the transformation rules (4.6) and (4.7) with respect to the elements of a reflection group G, is a* (–1,1)–differential of the group G. *If, in addition,* $\|\mu\|_\infty < 1$, *then μ is a* Beltrami differential of the group G. *We use the notation $D^{(-1,1)}(G)$ for (–1,1)–differentials of a reflection group G, and the notation $Bel(G)$ for Beltrami differentials of G.*

Provided that the group G does not contain orientation reversing elements $D^{(-1,1)}(G)$ is a complex Banach space and $Bel(G)$ is its open unit ball. If G contains also glide–reflections, $D^{(-1,1)}(G)$ is a real Banach space.

There is a real analytic homeomorphism between the unit disk and the complex plane. Nevertheless, there are no quasiconformal mappings between them. This follows rather easily from the quasi–invariance of conformal invariants under quasiconformal mappings. The case of compact Klein surfaces is, however, different: homeomorphic compact Klein surfaces are also quasiconformally equivalent by the following result.

Theorem 4.2.2 *Let X and Y be compact Klein surfaces. Each homeomorphism $f : X \to Y$ is homotopic to a quasiconformal mapping.*

Proof. It is well known that each homeomorphism $f : X \to Y$ is homotopic to a diffeomorphism. The dilatation quotient of such a diffeomorphism can be computed at each point of X. It does not depend on the choices of the local variables, and it is a continuous function on X. Since X is compact, this function has a finite maximum K on X. Hence a diffeomorphism homotopic to f is a quasiconformal mapping of X. There are many ways to find a diffeomorphism homotopic to a given homeomorphism. For more details and a direct construction we refer to [60, Theorem V.1.5.].

Let X and Y be homeomorphic compact Klein surfaces. By Theorem 4.2.2 there are quasiconformal mappings $X \to Y$. Teichmüller considered the problem of finding, in a given homotopy class of mappings $X \to Y$, one with the smallest maximal dilatation. He gave a complete solution to this problem. He showed the following result.

Theorem 4.2.3 (Teichmüller extremal mapping theorem) *Let X and Y be compact Riemann surfaces of genus $g > 1$ and $f : X \to Y$ a homeomorphism. There exists always a unique quasiconformal mapping $F : X \to Y$ such that the following holds:*

- *F is homotopic to the mapping f.*
- *If $g : X \to Y$ is a K_g-quasiconformal mapping and homotopic to f, then $K_g \geq K_f$, where K_F is the maximal dilatation of F. If $K_g = K_F$, then also $g = F$.*

Teichmüller gave also an explicit description of the geometry of such an extremal mapping F. The most delicate part of this result is the uniqueness of the extremal mapping. Detailed proofs for the above results of Teichmüller can be found, for instance, in [60, Chapter V.]. Observe that non–classical surfaces are not usually considered in this context but the above result holds also for them. In other words we have: *A homotopy class of a homeomorphism between compact non–classical compact Riemann surfaces of genus $g > 1$ always contains a unique quasiconformal mapping having the smallest maximal dilatation.*

It is out of the scope of the present monograph to prove these results. A clear exposition can be found, for instance, in the monograph of Olli Lehto [60]. Observe that, in the above theorems, the uniqueness part is not anymore true if we drop the assumption $g > 1$.

4.3 Teichmüller spaces of Klein surfaces

There are many alternative definitions for the Teichmüller space. In these sections we will discuss some of these equivalent definitions.

In certain cases it is convenient to consider C^∞–surfaces Σ instead of topological surfaces. In regard of the moduli problem this does not imply any restriction of generality since a topological surface always has an infinitely differentiable structure and any two such structures of a surface are isomorphic (as C^∞–structures). So we feel free to make, at this point, the assumption that Σ is a C^∞–surface rather than just a topological surface.

Let Σ be a fixed compact C^∞–surface. The surface Σ need not be orientable, and it may have a non–empty boundary. If Σ is orientable, then we assume also that it *is* oriented. For an oriented surface Σ, let $\mathcal{M}(\Sigma)$ denote the set of those complex structures of Σ which *agree with the orientation and the differentiable structure.* For a non–orientable surface Σ the set $\mathcal{M}(\Sigma)$ consists of dianalytic structures of Σ that agree with the differentiable structure.

Next we consider the group $\mathrm{Diff}(\Sigma)$ consisting of diffeomorphic self–mappings of Σ. $\mathrm{Diff}(\Sigma)$ acts on $\mathcal{M}(\Sigma)$ in the following way. Let $f : \Sigma \to \Sigma$ be a diffeomorphism and $X \in \mathcal{M}(\Sigma)$ a complex structure of Σ. The complex structure $f^*(X)$ of Σ is defined by requiring the mapping

$$f : (\Sigma, f^*(X)) \to (\Sigma, X) \tag{4.8}$$

be dianalytic. More precisely, if Σ is oriented and f orientation preserving, then the mapping (4.8) has to be holomorphic. For orientation reversing mappings f, (4.8) is antiholomorphic. If Σ is non–orientable, then the mapping (4.8) has to be dianalytic. It is immediate that, in all cases, these conditions determine uniquely the analytic structure $f^*(X) \in \mathcal{M}(\Sigma)$ of Σ. This is how a diffeomorphism $f : \Sigma \to \Sigma$ induces a mapping $f^* : \mathcal{M}(\Sigma) \to \mathcal{M}(\Sigma), X \mapsto f^*(X)$.

Use the notation

$$\mathrm{Diff}_0(\Sigma) = \{f \in \mathrm{Diff}(\Sigma) \mid f \text{ homotopic to the identity}\}.$$

The group $\mathrm{Diff}_0(\Sigma)$ acts freely on $\mathcal{M}(\Sigma)$. That fact is one justification for the following definition.

Definition 4.3.1 (of Teichmüller spaces) *The set theoretic quotient*

$$T(\Sigma) = \mathcal{M}(\Sigma)/\mathrm{Diff}_o(\Sigma)$$

is the Teichmüller space *of the surface Σ. We use also the notation T^g for the Teichmüller space of a genus g surface Σ.*

Observe that homotopic self–mappings of an oriented surface are simultaneously orientation preserving. Consequently, if Σ is oriented, then $\mathrm{Diff}_0(\Sigma)$ is a subgroup of the group $\mathrm{Diff}_+(\Sigma)$ of orientation preserving diffeomorphic self–mappings of Σ.

Definition 4.3.2 (of the modular group) *For an oriented surface Σ, the group $\Gamma(\Sigma) = \mathrm{Diff}_+(\Sigma)/\mathrm{Diff}_0(\Sigma)$ is the* modular group *or the* mapping class group *of the surface Σ. For non-orientable surfaces Σ* the modular group *is defined setting $\Gamma(\Sigma) = \mathrm{Diff}(\Sigma)/\mathrm{Diff}_0(\Sigma)$. We use also the notation Γ^g for the modular group of a genus g surface Σ.*

Definition 4.3.3 (of the moduli space) *For an oriented surface Σ, the quotient $M(\Sigma) = \mathcal{M}(\Sigma)/\mathrm{Diff}_+(\Sigma)$ is the* moduli space . The moduli space *of a non-orientable surface Σ is $M(\Sigma) = \mathcal{M}(\Sigma)/\mathrm{Diff}(\Sigma)$. We use also the notation M^g for the moduli space of a smooth genus g surface Σ.*

It follows from the above definitions that the modular group acts on the Teichmüller space, and that $M(\Sigma) = T(\Sigma)/\Gamma(\Sigma)$. An orientation reversing self-mapping f of Σ induces, likewise, a mapping $f^* : T(\Sigma) \to T(\Sigma), [X] \mapsto [f^*(X)]$, where the analytic structure $f^*(X)$ of Σ is defined requiring (4.8) be analytic.

The above definitions could as well have been given in terms of hyperbolic metrics of Σ. In the analytic definition of the Teichmüller space we identify two analytic structures of a surface Σ if the homotopy class of the identity mapping of Σ contains a dianalytic mapping. In the geometric definition for Teichmüller spaces we identify two hyperbolic metrics of a given surface Σ if the homotopy class of the identity mapping contains an isometry between these two metrics.

Let $X, Y \in \mathcal{M}(\Sigma)$ be two analytic structures of a fixed compact surface Σ. By Theorem 4.2.2 there are quasiconformal mappings $(\Sigma, X) \to (\Sigma, Y)$ homotopic to the identity mapping of Σ. For such a quasiconformal mapping f, let K_f denote the maximal dilatation of f.

Definition 4.3.4 (of the Teichmüller metric) *The distance between two points $[X]$ and $[Y]$ of the Teichmüller space $T(\Sigma)$ in* the Teichmüller metric *τ of the $T(\Sigma)$ is defined by*

$$\tau([X],[Y]) = \inf\{\frac{1}{2}\log K_f | \, f \in \mathrm{Diff}_0(\Sigma)\}.$$

The Teichmüller space $T(\Sigma)$ together with the Teichmüller metric is homeomorphic to an Euclidean space ([60, Theorem 9.2, page 241]). This is one of the deep consequences of the Teichmüller extremal mapping theorem that cannot be presented in this monograph. We refer to Chapter V of the monograph of Olli Lehto ([60]) for a complete and detailed treatment of this aspect of the theory of Teichmüller spaces.

We observe, nevertheless, that the elements of the modular group $\Gamma(\Sigma)$ are isometries of the Teichmüller metric. This is an immediate consequence of the definitions. Hence they are, in particular, homeomorphic self-mappings of the Teichmüller space.

4.4 Teichmüller spaces of Beltrami differentials

By Theorem 4.2.1 we can associate, to each Beltrami differential μ of a Klein surface X, a Klein surface X_μ. Consequently, Teichmüller spaces can be defined also in terms of Beltrami differentials.

To be more precise, choose a point $[X] \in T(\Sigma)$, which we will refer to as *the origin* of the Teichmüller space. Consider the space $Bel(X)$ of Beltrami differentials of X. Each $\mu \in Bel(X)$ defines a unique $X_\mu \in \mathcal{M}(\Sigma)$ such that the identity mapping of Σ is a μ– quasiconformal mapping $X \to X_\mu$. We say that two Beltrami differentials μ_1 and μ_2 of X are *equivalent*, $\mu_1 \approx \mu_2$, if the homotopy class of the identity mapping of Σ contains an isomorphism $(\Sigma, X_{\mu_1}) \to (\Sigma, X_{\mu_2})$.

Definition 4.4.1 *The set* $T(X) = Bel(X)/\approx$ *is* the Teichmüller space of Beltrami differentials of X.

It is an immediate consequence of these definitions that

$$T(X) \to T(\Sigma), [\mu] \mapsto [X_\mu], \tag{4.9}$$

is a bijection between these two Teichmüller spaces.

The Teichmüller metric of $T(X)$ is the pull back of the Teichmüller metric of $T(\Sigma)$ under the mapping (4.9).

Theorem 4.4.1 *Teichmüller space* $T(X)$ *is connected.*

Proof. Let μ represent an arbitrary point of $T(X)$. Then $t \mapsto [t\mu]$ is a path connecting the point $[\mu]$ to the origin of $T(X)$. Note that this path depends on the choice of μ ([102]).

Lars Ahlfors showed as early as in 1959([5]) that, for an orientable surface Σ without boundary, the Teichmüller space $T(\Sigma)$ has a natural complex structure and is a complex manifold. The construction of Ahlfors was based on considering periods of Abelian differentials. It is not possible to present it here. We will, however, describe a way to decide which functions are holomorphic on $T(\Sigma)$. *Let us assume now that* Σ *is a compact and oriented surface without boundary.*

It is convenient to consider the Teichmüller space $T(X)$ of Beltrami differentials of X instead of $T(\Sigma)$. Let $\pi : Bel(X) \to T(X)$ be the projection. This is a continuous mapping with respect to the L^∞–metric on $Bel(X)$ and the Teichmüller metric on $T(X)$ (cf. e.g. [60, III.2.2]). Hence, for any open $U \subset T(X)$, $\pi^{-1}(U)$ is open in $Bel(X)$.

Holomorphic functions on $T(X)$. Let $U \subset T(X)$ be open. We declare a function $f : U \to \mathbf{C}$ *holomorphic* if the composition $f \circ \pi$ is a holomorphic function on the open set $\pi^{-1}(U)$ of the complex Banach space of (–1,1)–differentials of X.

In this way $T(X)$ becomes first *a ringed space*. It is not, *a priori,* clear that the above definition actually gives a good complex structure on $T(X)$. That is, however, the case if Σ is an oriented surface which does not have boundary (cf. e.g. [60, Chapter V]). This complex structure of $T(X)$ is then transformed to a complex structure of $T(\Sigma)$ requiring the mapping (4.9) be holomorphic.

We still have to check that this complex structure of $T(\Sigma)$ does not depend on the choice of the origin X of the Teichmüller space $T(\Sigma)$. But that is an immediate consequence of the transformation formula (2.8) and the remark made after it.

It is a rather straightforward verification that the elements of the modular group are biholomorphic automorphisms of the Teichmüller space. *Royden has shown*[1] *([71, Theorem 1 on page 281] and [72, Theorem 2 on page 379]), in fact, that for surfaces of genus $g > 2$, $\Gamma(\Sigma)$ is the full group of holomorphic automorphisms of $T(\Sigma)$.* In the same way we verify that the mapping $\sigma^* : T(\Sigma) \to T(\Sigma)$, induced by an orientation reversing mapping $\sigma : \Sigma \to \Sigma$ (cf. formula (4.8)), is an antiholomorphic self–mapping of the Teichmüller space (for details see [74, 5.10]).

For a later reference we formulate these results as follows:

Theorem 4.4.2 *The Teichmüller space of compact genus g Riemann surfaces is a complex manifold of complex dimension $3g - 3$. The elements of the modular group Γ^g are holomorphic automorphisms of T^g. For $g > 2$, Γ^g is the full group of holomorphic automorphisms of T^g.*

This result can be extended to general finite dimensional Teichmüller spaces of oriented surfaces without boundaries. For a careful proof of this result we refer to the monograph of F. W. Gardiner [32, 9.2].

4.5 Non–classical Klein surfaces

If Σ has a non–empty boundary or if it is non–orientable, then $T(\Sigma)$ is *not* a complex manifold. In this case the Teichmüller space is, in a natural way, a real analytic manifold. This real analytic structure can be obtained from the complex structure of Teichmüller spaces compact Riemann surfaces. To this end a topological construction is necessary.

[1] Clifford Earle has recently found a new proof for this result of Royden.

Assume that Σ is a compact surface which either is not orientable or has a non-empty boundary (or both). Recall the definition of *the complex double*, Σ^c, of the surface Σ (see page 77).

Σ^c is a compact oriented surface without boundary together with a projection $\pi : \Sigma^c \to \Sigma$ that is a ramified double covering mapping. It is ramified precisely at the points lying over the boundary points of Σ. The covering group of π is generated by an orientation reversing involution σ of Σ^c for which $\pi \circ \sigma = \pi$. The fixed point set of the involution σ corresponds to the boundary points of Σ.

Above we have observed that the mapping $\sigma : \Sigma^c \to \Sigma^c$ induces an antiholomorphic self-mapping σ^* of $T(\Sigma^c)$. This mapping is an involution since σ is an involution.

For any $X \in \mathcal{M}(\Sigma)$ let $\pi^*(X)$ be the complex structure of Σ^c which agrees with the orientation of Σ^c and for which the projection

$$\pi : (\Sigma^c, \pi^*(X)) \to (\Sigma, X)$$

is dianalytic. It is immediate that

$$\pi^* : T(\Sigma) \to T(\Sigma^c), [X] \mapsto [\pi^*(X)],$$

is a well-defined mapping of Teichmüller spaces. It is not difficult to show that the mapping $\pi^* : T(\Sigma) \to T(\Sigma^c)$ is an isometry with respect to the corresponding Teichmüller metrics. Hence it is, in particular, a homeomorphism of $T(\Sigma)$ onto $\pi^*(T(\Sigma))$.

Theorem 4.5.1 *Assume that Σ is a non-classical surface of genus g, $g > 1$. We have*

$$\pi^*(T(\Sigma)) = T(\Sigma^c)_{\sigma^*}$$

Proof. Let X be a complex structure of Σ. The complex structure $\pi^*(X)$ of Σ^c has the (defining) property that $\sigma : (\Sigma^c, \pi^*(X)) \to (\Sigma^c, \pi^*(X))$ is an antiholomorphic involution. This implies that

$$[\pi^*(X)] \in T(\Sigma^c)_{\sigma^*}, \text{ i.e. that } \pi^*(T(\Sigma) \subset T(\Sigma^c)_{\sigma^*}.$$

To prove the converse inclusion, take a point $[Y] \in T(\Sigma^c)_{\sigma^*}$. By the definition this means that there is a holomorphic mapping $f : (\Sigma^c, Y) \to (\Sigma^c, \sigma^*(Y))$ which is homotopic to the identity mapping of Σ^c. (Recall that $\sigma^*(Y)$ is defined as that complex structure of Σ^C for which the mapping $\sigma : (\Sigma^c, \sigma^*(Y)) \to (\Sigma^c, Y)$ is antiholomorphic.)

The construction implies that $\tau = \sigma \circ f : (\Sigma^c, Y) \to (\Sigma^c, Y)$ is an antiholomorphic mapping. Then $\tau^2 : (\Sigma^c, Y) \to (\Sigma^c, Y)$ is a holomorphic mapping. Since σ is an involution and f is homotopic to the identity, we conclude that τ is also homotopic to the identity. Since the genus of the Riemann surface (Σ^c, Y) is at least 2, the only holomorphic automorphism of (Σ^c, Y) that is homotopic to the identity is, by Lemma 3.6.8, the identity itself. This implies that $\tau : (\Sigma^c, Y) \to (\Sigma^c, Y)$ is an antiholomorphic *involution* proving the theorem.

For a surface Σ that is either non–orientable or has a non–empty boundary (or both) we now identify $T(\Sigma)$ with its image $\pi^*(T(\Sigma)) = T(\Sigma^c)_{\sigma^*}$ in $T(\Sigma^c)$. Since σ^* is an antiholomorphic involution of $T(\Sigma^c)$, its fixed–point set, $\pi^*(T(\Sigma))$, is a real analytic manifold.

4.6 Teichmüller spaces of genus 1 surfaces

Classical and non–classical compact genus 1 Riemann surfaces are:

- torus,
- annulus,
- Klein bottle,
- Möbius band.

Last three are non–classical and their complex double, in all cases, is the torus. By Theorem 4.5.1 we can interpret the Teichmüller space of these non–classical surfaces as a subset of the Teichmüller space of the torus. In this case the Teichmüller space is quite well known. To motivate later constructions we will, in this Section, review the classical theory concerning the Teichmüller spaces of genus 1 surfaces.

Recall considerations related to the Riemann Mapping Theorem 3.6.1 on page 86: the universal covering surface of a torus T^1 is the finite complex plane $\mathbf{C}$. The covering group G is generated by two translations $g_1(z) = z + \omega_1$ and $g_2(z) = z + \omega_2$. Let $\tau = \omega_1/\omega_2$. The quotient $\mathbf{C}/G$ is a compact genus 1 surface if and only if τ is not real. Without loss of generality we may then assume that $\operatorname{Im} \tau > 0$. This representation for a torus is called *normalized.*

Let T^1 be a *topological* oriented torus, and let X be a complex structure on it which agrees with the orientation. The fundamental group $\pi_1(T^1, P)$ of T^1 at a base–point $P \in T^1$ is generated by homotopy classes of two simple closed curves α and β (see Fig. 4.2).

Let X and X' be two complex structures of T^1 which agree with the orientation. Let $(T^1, X) = \mathbf{C}/G$ and $(T^1, Y) = \mathbf{C}/G'$, where G is generated

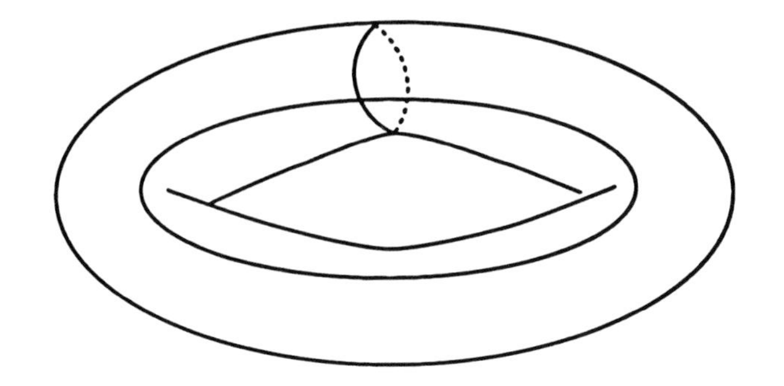

Figure 4.2: The fundamental group of a torus is generated by the homotopy classes of two simple closed curves.

by $z \mapsto z + \omega_1$, $z \mapsto z + \omega_2$ and G' is generated by $z \mapsto z + \omega_1'$, $z \mapsto z + \omega_2'$. Without loss of generality we may assume that $z \mapsto z + \omega_1$ and $z \mapsto z + \omega_1'$ both cover the homotopy class of α and $z \mapsto z + \omega_2$, $z \mapsto z + \omega_2'$ cover that of β. This means that the identity mapping $(T^1, X) \to (T^1, X')$ induces the isomorphism $G \to G'$ which takes $z \mapsto z + \omega_1$ onto $z \mapsto z + \omega_1'$ and $z \mapsto z + \omega_2$ onto $z \mapsto z + \omega_2'$.

Assume now that X and X' determine the same point in the Teichmüller space $T(T^1)$ of the torus T^1. This means that there is a holomorphic homeomorphism $f : (T^1, X) \to (T^1, X')$ which is homotopic to the identity mapping of T^1. By Lemma 3.6.7 (page 92) this means that f has a lifting $F : \mathbf{C} \to \mathbf{C}$ which induces the same isomorphisms as the identity mapping.

The mapping F is now of the form $F(z) = az + b$, where $a \neq 0$. The condition that it induces the same isomorphism as the identity mapping simply means that

$$F(z + \omega_1) = F(z) + \omega_1' \text{ and } F(z + \omega_2) = F(z) + \omega_2' \tag{4.10}$$

Since $a \neq 0$, equations (4.10) yield

$$\frac{\omega_1}{\omega_2} = \frac{\omega_1'}{\omega_2'}.$$

This means that the mapping

$$T(T^1) \to U, \ [X] \mapsto \frac{\omega_1}{\omega_2} =: \tau \tag{4.11}$$

is a well defined mapping. Reversing the above argument shows that it is also injective and onto. We have, therefore:

Lemma 4.6.1 *The Teichmüller space $T(T^1)$ of a torus T^1 is the upper half-plane U.*

By the criterium characterizing the complex structure of $T(T^1)$ (on page 148) it is immediate that the mapping (4.11) is a holomorphic homeomorphism. We may, therefore, identify the Teichmüller space of a torus by the upper half-plane.

Let Σ be a non-classical compact Riemann surface of genus 1. Then Σ may be expressed as a quotient

$$\Sigma = T^1/\langle\sigma\rangle,$$

where $\sigma : T^1 \to T^1$ is an orientation reversing involution. The mapping $\sigma : T^1 \to T^1$ induces an antiholomorphic involution $\sigma^* : T(T^1) \to T(T^1)$, i.e., an antiholomorphic involution $\sigma^* : U \to U$ of the upper half-plane.

A closer analysis reveals that we may assume that σ^* is one of the following involutions of the upper half-plane:

1. $\sigma^*(z) = -\overline{z}$.

2. $\sigma^*(z) = 1/\overline{z}$.

3. $\sigma^*(z) = -\overline{z} + 1$.

The first involution corresponds to the annulus and to the Klein bottle. The two following involutions correspond to the Möbius strip. Here we have the phenomena that the images of Teichmüller spaces of two non-classical genus 1 surfaces Σ and Σ' agree in the Teichmüller space of their common complex double, namely the torus. This does not happen in the case of Teichmüller spaces of surfaces of genus g, $g > 2$.

4.7 Teichmüller spaces of reflection groups

Points of the Teichmüller space of a surface Σ are equivalence classes of dianalytic structures on Σ. Choosing an origin $X = (\Sigma, X)$ for the Teichmüller space, we can view its points as quasiconformal deformations of the Klein surface X corresponding to the origin.

Via the Uniformization this can be transformed to groups. We use the standard notations: $GL(2,\mathbf{R})$ is the group of real 2×2–matrices with determinant $\neq 0$, $PGL(2,\mathbf{R})$ is the group of holomorphic or antiholomorphic automorphisms of the upper half-plane, $SL(2,\mathbf{R})$ is the group of real 2×2–matrices with determinant $+1$, and $PSL(2,\mathbf{R})$ is group of Möbius–transformations mapping U onto itself. Every reflection group G is a discrete subgroup of $PGL(2,\mathbf{R})$.

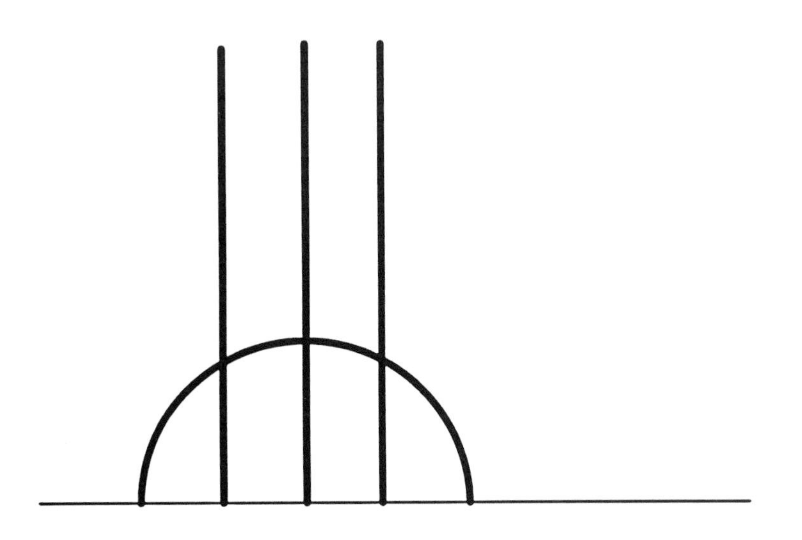

Figure 4.3: Teichmüller spaces of non–classical genus 1 Riemann surfaces as subsets of the Teichmüller space of a torus.

Assume now that $G \subset PGL(2,\mathbf{R})$ is a fixed reflection group representing the origin $[X]$ of the Teichmüller space.

Definition 4.7.1 The deformation space $\mathcal{R}^0(G, PGL(2,\mathbf{R})) = \mathcal{R}^0(G)$ *of* G *in* $PGL(2,\mathbf{R})$ *is the space of all injective homomorphisms* $\theta : G \to PGL(2,\mathbf{R})$ *such that* $\theta(G)$ *is discrete and the Klein surfaces* U/G *and* $U/\theta(G)$ *are quasiconformally equivalent.*

The condition about quasiconformal equivalence simply means that the Klein surfaces U/G and $U/\theta(G)$ have to be of the same type, for instance homeomorphic compact surfaces or interiors of mutually homeomorphic compact Klein surfaces with boundary.

The deformation space $\mathcal{R}^0(G)$ is a subset of the space $\mathcal{R}(G)$ of all homomorphisms of G into $PGL(2,\mathbf{R})$. Both these spaces are subspaces of $PGL(2,\mathbf{R})^G$ which carries the product topology. This topology induces a topology on $\mathcal{R}^0(G)$ and on $\mathcal{R}(G)$.

Every isomorphism $\theta : G \to \theta(G) \subset PGL(2,\mathbf{R})$ defines a Klein surface $X_\theta = U/\theta(G)$ which is homeomorphic to $X = U/G$. By a theorem of Nielsen all such isomorphisms θ are *geometric*, i.e., there exists a homeomorphism $\varphi : X \to X_\theta$ which induces the isomorphism θ.

Definition 4.7.2 *Two isomorphisms* $\theta_j : G \to \theta_j(G)$, $j = 1, 2$, *are* equivalent, $\theta_1 \approx \theta_2$, *if* $\theta_1 \circ \theta_2^{-1}$ *is the restriction to* $\theta_2(G)$ *of an inner automorphism* *of* $PGL(2,\mathbf{R})$. *The quotient* $T(G) = \mathcal{R}^0(G)/\approx$ *is* the Teichmüller space *of the group* G.

Recall that we started with a surface Σ of genus g, $g > 1$, and chose an origin $X = (\Sigma, X)$ for the Teichmüller space and fixed a reflection group G such that $X = U/G$.

Let $\theta : G \to \theta(G)$ be an isomorphism representing a point in $T(G)$. The above mentioned Theorem of Nielsen together with Theorem 4.2.2 implies that there is a quasiconformal mapping $f : X \to U/\theta(G)$ which induces the isomorphism θ in the sense of Definition 3.6.3. Let μ be the complex dilatation of f. By Theorem 4.2.1 the Beltrami–differential μ defines a complex structure X_μ of Σ such that the identity mapping $(\Sigma, X) \to (\Sigma, X_\mu)$ is μ–quasiconformal.

Theorem 4.7.1 *The formula*

$$\Lambda : T(G) \to T(\Sigma), [\theta] \mapsto [X_\mu]$$

defines a bijective mapping between the two Teichmüller spaces.

Proof. Let us first show that Λ is a well defined mapping. To that end let θ_1 and θ_2 be equivalent isomorphisms of G. Let $f_j : X \to U/\theta_j(G)$ be a μ_j quasiconformal mapping inducing θ_j, $j = 1, 2$. We have to show that the homotopy class of the identity mapping of Σ contains a dianalytic mapping $(\Sigma, X_{\mu_1}) \to (\Sigma, X_{\mu_2})$.

Since θ_1 is equivalent to θ_2 there exists, by the definition, an element $t \in PGL(2,\mathbf{R})$ such that

$$\theta_1(g) = t \circ \theta_2(g) \circ t^{-1}$$

holds for all $g \in G$. By Lemma 3.6.7 we conclude that $f_1 \circ f_2^{-1}$ is homotopic to a mapping $h : U/\theta_2(G) \to U/\theta_1(G)$ which is either conformal or anticonformal.

We need more notation. For $j = 1, 2$, let $I_j : (\Sigma, X) \to (\Sigma, X_{\mu_j})$ be the identity mapping. These mappings form the following diagram.

$$\begin{array}{ccccc} & & X_{\mu_2} & & \\ & & \uparrow I_2 & & \\ X_{\mu_1} & \xleftarrow{I_1} & X & \xrightarrow{f_2} & U/\theta_2(G) \\ & & \downarrow f_1 & & \\ & & U/\theta_1(G) & & \end{array} \tag{4.12}$$

Consider the mapping

$$\alpha = I_1 \circ f_1^{-1} \circ h \circ f_2 \circ I_2^{-1} : (\Sigma, X_{\mu_2}) \to (\Sigma, X_{\mu_1}).$$

Since $h \approx f_1 \circ f_2^{-1}$, $f_1^{-1} \circ h \circ f_2$ is homotopic to the identity mapping of Σ. Consequently, α is homotopic to the identity as well. Since $I_j : (\Sigma, X) \to (\Sigma, X_{\mu_j})$ and $f_j : (\Sigma, X) \to U/\theta_j(G)$ are both μ_j- quasiconformal, $j = 1, 2$, the mappings $I_1 \circ f_1^{-1}$ and $f_2 \circ I_2^{-1}$ are both conformal by the uniqueness of quasiconformal mappings (cf. page 65). It follows that $\alpha : (\Sigma, X_{\mu_2}) \to (\Sigma, X_{\mu_1})$ is a conformal or anticonformal mapping homotopic to the identity. This proves that the mapping Λ is well defined.

Let us next prove that Λ is injective. To that end suppose that $\theta_j : G \to \theta_j(G)$, $j = 1, 2$, are two isomorphisms such that $\Lambda([\theta_1]) = \Lambda([\theta_2])$. Let $f_j : X \to U/\theta_j(G)$ be the μ_j- quasiconformal mapping inducing θ_j. As above, let $I_j : (\Sigma, X) \to (\Sigma, X_{\mu_j})$ be the identity mapping of Σ. These mappings form again the diagram (4.12). Since $\Lambda([\theta_1]) = \Lambda([\theta_2])$, X_{μ_1} and X_{μ_2} define the same point in the Teichmüller space. Hence the homotopy class of the identity mapping contains a dianalytic mapping $\alpha : (\Sigma, X_{\mu_2}) \to (\Sigma, X_{\mu_1})$. Consider the mapping

$$f_1 \circ I_1^{-1} \circ \alpha \circ I_2 \circ f_2^{-1} : U/\theta_2(G) \to U/\theta_1(G) \tag{4.13}$$

By the uniqueness of quasiconformal mappings, the mapping (4.13) is dianalytic. Its lifting to the upper half-plane is a dianalytic mapping which conjugates $\theta_1(G)$ to $\theta_2(G)$. This argument shows that the isomorphisms $\theta_1 : G \to \theta_1(G)$ and $\theta_2 : G \to \theta_2(G)$ are equivalent. Hence the mapping Λ is injective.

We still have to show that Λ is surjective. To that end, let $[Y] \in T(\Sigma)$ be an arbitrary point of $T(\Sigma)$. Then there exists a Beltrami differential μ of X such that $[X_\mu] = [Y]$. This Beltrami differential lifts to a Beltrami differential of the reflection group G, $X = U/G$. We denote the lifting of μ also by the same symbol. Let $f_\mu : U \to U$ be a μ-quasiconformal self mapping of the upper half-plane. Then $\theta : G \to \theta(G) \subset PGL(2, \mathbf{R}), g \mapsto f_\mu \circ g \circ f_\mu^{-1}$ determines a point in $T(G)$ which maps to $[Y]$ under the mapping Λ. Hence Λ is surjective.

4.8 Parametrization of Teichmüller spaces

We will now apply the machinery developed in Chapter 1 to Teichmüller spaces. Let us first consider orientable surfaces and their Teichmüller spaces only. *So we assume now that Σ is an orientable surface without boundary.* Choose an origin $[X]$ for the Teichmüller space and a Fuchsian group G such that $X = U/G$.

Since Σ is orientable we may assume that $G \subset PSL(2, \mathbf{R})$ and consider an isomorphism $\theta : G \to \theta(G) \subset PGL(2, \mathbf{R})$. The group $\theta(G)$ acts on the upper half-plane — its elements are holomorphic or antiholomorphic automorphisms of U. Assume that $\theta(G)$ contains both holomorphic (other than the identity) and antiholomorphic automorphisms of U, i.e., that $\theta(G)$ is a proper reflection group. We show that this is not possible.

The elements of G, save the identity, do not have fixed-points in U. By Nielsen's theorem $\theta(G)$ is a geometric deformation of G. Hence the elements of $\theta(G)$ do not have fixed-points in U either.

We assumed that $\theta(G)$ is a proper reflection group. The above remark implies then that $U/\theta(G)$ is not orientable. But that is not possible since $U/\theta(G)$ is homeomorphic to U/G which is orientable.

This argument shows that as self-mappings of U the non-identity elements of $\theta(G)$ are all either holomorphic or antiholomorphic. But a group $\theta(G) \in PGL(2,\mathbf{R})$ containing antiholomorphic self-mappings of U contains also holomorphic self-mappings of U. We conclude that when starting with an orientable surface Σ each representation $\theta : G \to PGL(2,\mathbf{R})$ satisfies $\theta(G) \in PSL(2,\mathbf{R})$.

This shows that when considering Teichmüller spaces of orientable surfaces we may replace $PGL(2,\mathbf{R})$ by $PSL(2,\mathbf{R})$ in the construction of $T(G)$. Let us do this.

We assume that the classes of all possible pairs (g,h) of transformations of G are known. This is actually a topological condition: the class of a pair (g,h) is determined by the corresponding pair of elements of the fundamental group of Σ.

Let $\theta : G \to PSL(2,\mathbf{R})$ be a deformation of the group G. Recall that, by the above remark, all points of $T(G)$ can be represented by deformations of G into $PSL(2,\mathbf{R})$.

The following result has been presented already in Chapter 1 (Theorem 1.4.2 on page 28). Here we give a new proof which applies the theory of quasiconformal mappings.

Lemma 4.8.1 *Let (g,h) be a pair of elements of G whose class is defined. Then also the class of the pair $(\theta(g),\theta(h))$ is defined and is the same as the class of the pair (g,h).*

Proof. The class of a pair of hyperbolic Möbius-transformations is defined whenever the transformations do not have fixed-points in common. It is clear that if g and h do not have common fixed-points then neither do $\theta(g)$ and $\theta(h)$. Hence the statement about the classes being defined is apparent.

Recall that, by Nielsen's theorem, every isomorphism $\theta : G \to \theta(G)$ is geometric, i.e., is induced by a quasiconformal mapping $U/G \to U/\theta(G)$. Let μ be the complex dilatation of that mapping. Lift it to a Beltrami differential μ of the group G. Let $f_\mu : U \to U$ be a μ-quasiconformal mapping of U onto itself normalized in such a way that f_μ fixes 0, 1 and ∞. Then the isomorphism θ is equivalent to $g \mapsto f_\mu \circ g \circ f_\mu^{-1}$.

The class of a pair of Möbius–transformations is not affected by conjugation. Hence the classes of the pairs

$$(\theta(g), \theta(h)) \text{ and } (g_\mu, h_\mu) = (f_\mu g f_\mu^{-1}, f_\mu h f_\mu^{-1})$$

are the same. Therefore it suffices to prove that the classes of (g, h) and (g_μ, h_μ) are the same.

For each t, $0 \leq t \leq 1$, consider the Beltrami–differential $t\mu$ of the group G. Let $f_t : U \to U$ be the $t\mu$–quasiconformal mapping of U onto itself normalized in such a way that it fixes 0, 1 and ∞. Then each f_t induces an isomorphism

$$\theta_t : G \to \theta_t(G), g \mapsto f_t \circ g \circ f_t^{-1}.$$

The group G acts freely on U. Hence also all the groups $\theta_t(G) = f_t G f_t^{-1}$ act freely on U. It follows that none of the groups $\theta_t(G)$ contain elliptic transformations. We conclude that each pair $(\theta_t(g), \theta_t(h))$ belongs either to $\mathcal{P}$ or to $\mathcal{H}$.

The fixed points and the multipliers of $\theta_t(g)$ and $\theta_t(h)$ are continuous functions of t. Clearly also $\theta_0(g) = g$ and $\theta_0(h) = h$. We conclude, by the definition of the classes $\mathcal{P}$ and $\mathcal{H}$ (see page 24), that the class of $\theta_t(g), \theta_t(h))$ equals that of (g, h) for each t, $0 \leq t \leq 1$. In particular, the class of $(\theta_1(g), \theta_1(h))$ is the same as the class of (g, h) proving the lemma.

Consider the set $\mathcal{K}$ generating G and satisfying the relation (3.8, page 89). It is a simple matter to check that $\mathcal{K}$ satisfies the technical conditions of Lemma 1.6.3 on page 44. Also any deformation $\theta(\mathcal{K}) = \{\theta(f) | f \in \mathcal{K}\}$ of $\mathcal{K}$ satisfies those conditions. We will apply the consideration of Section 1.6 to this set $\mathcal{K}$.

Let $f_1, \ldots, f_{6p-4}$ be the elements of G whose multipliers determine $\mathcal{K}$ up to a conjugation by a Möbius–transformation. We can find these elements by Lemma 1.6.3 (page 44).

Recall that we have fixed an isomorphism $\pi_1(\Sigma, Q) \to G$. Let

$$\alpha_1, \ldots, \alpha_{6p-4}$$

be the closed curves of Σ corresponding to the elements $f_1, \ldots, f_{6p-4}$.

Let

$$\mathcal{L} : T(G) \to \mathbf{R}_+^{6p-4}, [G \xrightarrow{\theta} \theta(G)] \mapsto (\log k(\theta(f_1)), \ldots, \log k(\theta(f_{6p-4}))).$$

We use here the notation of Chapter 1: $k(\theta(f_j))$ is the multiplier of the Möbius–transformation $\theta(f_j)$. Lemma 1.6.3 (page 44) implies now:

Theorem 4.8.2 *The mapping*

$$\mathcal{L} : T(G) \to \mathbf{R}_+^{6p-4}$$

is injective.

4.9 Geodesic length functions

It is worthwhile to rephrase the above deliberations for the Teichmüller space $T(\Sigma)$ of a fixed surface Σ. Above we considered only the case of Fuchsian groups or the case of an oriented and compact surface Σ without boundary. That was done for mere technical convenience. Everything said here applies also to the more general case of surfaces of finite type.

Let α be a closed curve on Σ representing an element of the fundamental group $\pi_1(\Sigma, Q)$ of Σ at a point $q \in \Sigma$. Let X be a complex structure of Σ. Use the notation

$$\ell_\alpha(X) = \inf\{\text{length of } \beta \text{ on } X \,|\, \beta \text{ homotopic to } \alpha\}.$$

It is straightforward to check that

$$\ell_\alpha : T(\Sigma) \to \mathbf{R}_+,\ [X] \mapsto \ell_\alpha(X), \tag{4.14}$$

is a well defined function on $T(\Sigma)$.

Definition 4.9.1 *Functions of type (4.14) are called* geodesic length functions.

We show next that geodesic length functions are continuous following the argument of [89, Theorem 4.1].

Consider hyperbolic transformations g and g' which are conjugate under an orientation preserving homeomorphism $f : \hat{\mathbf{C}} \to \hat{\mathbf{C}}$, i.e.,

$$g' = f \circ g \circ f^{-1}.$$

If f is conformal, then $k(g') = k(g)$. We show next that the multiplier $k(g)$ is quasi-invariant under quasiconformal mappings.

Lemma 4.9.1 *If f is K-quasiconformal and $g' = f \circ g \circ f^{-1}$, then*

$$k(g)^{1/K} \leq k(g') \leq k(g)^K. \tag{4.15}$$

Proof. Denote $k = k(g)$ and $k' = k(g')$. We may assume that $g(z) = kz$ and $g'(z) = k'z$. Since $(g')^n = f \circ g^n \circ f^{-1}$, we have

$$f(k^n z) = (k')^n f(z) \tag{4.16}$$

for $n = 0, \pm 1, \pm 2, \ldots$. Let B_n be the annulus bounded by the circles $|z| = 1$ and $|z| = k^n$, $n = 1, 2, \ldots$. We approximate the ring domain $B'_n = f(B_n)$ by an annulus B''_n as follows. Let

$$\xi_1 = \min_{\vartheta} |f(e^{i\vartheta})| \text{ and } \xi_2 = \max_{\vartheta} |f(e^{i\vartheta})|.$$

Then $\xi_1 > 0$ and $\xi_2 < \infty$ and we set

$$B''_n = \{z | \xi_1 < |z| < (k')^n \xi_2\}.$$

Then by (4.16), $B'_n \subset B''_n$. Since, B'_n separates the boundary components of B''_n, we have

$$\begin{aligned} M(B'_n) &\leq M(B''_n) = \log \tfrac{(k')^n \xi_2}{\xi_1} \\ &= n \log k' + \log \tfrac{\xi_2}{\xi_1}. \end{aligned}$$

Here $M(B)$ denotes the modulus of the ring domain B as defined by Definition 2.2.4 (page 60).

On the other hand, since f is K-quasiconformal,

$$M(B'_n) \geq \frac{M(B_n)}{K} = \frac{n \log k}{K}.$$

So we have

$$\log k \leq K \log k' + \frac{K}{n} \log \frac{\xi_2}{\xi_1},$$

and, letting $n \to \infty$, we conclude that $\log k \leq K \log k'$. Similarly, replacing f by f^{-1}, we can conclude that $\log k' \leq K \log k$.

Theorem 4.9.2 *Geodesic length functions are continuous with respect to the topology induced by the Teichmüller metric.*

Proof. Let X be a complex structure on Σ representing a point in the Teichmüller space. Let α be a closed geodesic curve on Σ that is not homotopic to a point. Consider the geodesic length function ℓ_α at the point $[X] \in T(\Sigma)$.

Write $X = U/G$ for some Fuchsian group G. Assume that the Möbius transformation $g \in G$ covers the homotopy class of α. Let $k = k(g)$ be the multiplier of g. By the considerations on page 107 related to Theorem 3.12.1

$$\ell_\alpha([X]) = \log k.$$

Let Y be another complex structure of Σ and let $(\Sigma, Y) = U/G'$ for a Fuchsian group G'. Let $f : (\Sigma, X) \to (\Sigma, Y)$ be a K–quasiconformal mapping homotopic to the identity mapping of Σ. Let $F : U \to U$ be a lifting of f. Since $g \in G$ covers the homotopy class of the curve α and since f is homotopic to the identity, also $F \circ g \circ F^{-1} \in G'$ covers the homotopy class of α.

Let $k = k(g)$ and $k' = k(g')$ be the corresponding multipliers. By Lemma 4.9.1 we have:

$$\frac{1}{K} \leq \frac{\log k}{\log k'} \leq K \tag{4.17}$$

Taking the logarithms we get, from (4.17),

$$|\log(\log k) - \log(\log k')| \leq \log K, \tag{4.18}$$

i.e.,

$$|\log \ell_\alpha([X]) - \log \ell_\alpha([Y])| \leq \log K. \tag{4.19}$$

Inequality (4.19) is valid for all K that are maximal dilatations of quasiconformal mappings $f : (\Sigma, X) \to (\Sigma, Y)$ homotopic to the identity. In view of the definition of the Teichmüller metric τ (Definition 4.3.4 on page 146) (4.19) implies that

$$|\log \ell_\alpha([X]) - \log \ell_\alpha([Y])| \leq 2\tau([X], [Y]).$$

The geodesic length function ℓ_α is therefore continuous.

Using the description of the complex structure of the Teichmüller space given on page 148, we can show even more:

Theorem 4.9.3 *Geodesic length functions are real analytic.*

Proof. Choose an origin X for the Teichmüller space $T(\Sigma)$ and a Fuchsian group G with $U/G = X$. Fix an isomorphism $\pi_1(\Sigma, Q) \to G$ and let $g \in G$ be the Möbius–transformation corresponding to α. As in the above,

$$\ell_\alpha(X) = \log k(g).$$

Let μ be a Beltrami differential representing an arbitrary point of $T(X)$ the Teichmüller space of Beltrami differentials on X.

In view of what is said about the complex structure of $T(X)$ it suffices to show that

$$Bel(X) \to \mathbf{R}_+, \ \mu \mapsto \ell_\alpha(X_\mu),$$

is a real analytic mapping of the open unit ball in the complex Banach space of $(-1, 1)$–differentials of X.

The identity mapping $I_\mu : (\Sigma, X) \to (\Sigma, X_\mu)$ is a μ–quasiconformal mapping. Lift the Beltrami–differential μ of X to a Beltrami–differential

μ of the Fuchsian group G. Let f_μ be the μ–quasiconformal self–mapping of U normalized in such a way that f_μ keeps 0, 1 and ∞ fixed. Then $G_\mu = f_\mu G(f_\mu)^{-1}$ is a Fuchsian group with $X_\mu = U/G_\mu$. The projection $U \to X_\mu$ being defined in such a way that the diagram

$$\begin{array}{ccccccc} & & G & & G_\mu & & \\ & & U & \xrightarrow{f_\mu} & U & & \\ & & \downarrow{\scriptstyle \pi} & & \downarrow & & \\ U/G & = & X & \xrightarrow{I_\mu} & X_\mu & = & U/G_\mu \end{array}$$

commutes. In other words, $U \to X_\mu$ is the universal covering of X_μ, the covering projection is $I_\mu \circ \pi \circ (f_\mu)^{-1}$ where $\pi : U \to X$ is the projection $U \to X$ and the covering group is $G_\mu = f_\mu G f_\mu^{-1}$. The element $g \in G$ covers the curve α on X. Consequently, the element $f_\mu \circ g \circ f_\mu^{-1}$ covers the curve α on X_μ. Then, repeating the above considerations related to Theorem 3.12.1,

$$\ell_\alpha(X_\mu) = \log k(f_\mu \circ g \circ (f_\mu)^{-1}). \tag{4.20}$$

The right–hand side of (4.20) depends real analytically on $\mu \in Bel(G)$. This is a consequence of an important result in the theory of quasiconformal mappings according to which a quasiconformal mapping depends holomorphically on its complex dilatation. Here it is not possible to go into the details in this matter. For a detailed discussion we refer to [60, § 3.2., pp. 69 – 72] and to references given there. Hence, in view of the definition of the complex structure of $T(\Sigma)$, ℓ_α is a real analytic function.

We conclude this chapter observing that identifying $T(\Sigma)$ with $T(G)$ the parametrization $\mathcal{L}$ for $T(G)$ is actually a parametrization of $T(\Sigma)$ by geodesic length functions. Furthermore, $\mathcal{L}$ is real analytic. Here we have considered only orientable compact surfaces without boundary. Surfaces with boundary and non–orientable surfaces can be treated by the above methods passing first to their complex doubles. We shall not do it here. For details see [80].

Consider the image $\mathcal{L}(T(\Sigma))$ of the Teichmüller space in $\mathbf{R}_+^{6p-4}$. Each point of $T(\Sigma)$ is an equivalence class of complex structures of Σ or an equivalence class of hyperbolic metrics. Let us now use the latter interpretation. It is of interest to try to see what happens to the metric as we approach the boundary of $\mathcal{L}(T(\Sigma))$ in $\mathbf{R}_+^{6p-4}$. Theorem 3.19.1 is our main tool in studying the possible ways of degeneration of Riemann surfaces. That will be the main topic of Chapter 5.

4.10 Discontinuity of the action of the modular group

In this Section we take a closer look at the action of the modular group on the Teichmüller space. We start with some rather obvious technical results.

Consider the Teichmüller space $T(\Sigma)$ of a compact surface of genus $g > 1$. Let $G \subset \Gamma(\Sigma)$ be a finite subgroup of the modular group.

Lemma 4.10.1 *If $[X] \in T(\Sigma)$ is fixed by all elements of G, then G is a quotient of the automorphism group of the Riemann surface X.*

Proof. This rather obvious by the definitions and by the fact that the homotopy class of a homeomorphic self–mapping of a compact Riemann surface of genus > 1 contains at most one automorphism.

We usually have $G = \mathrm{Aut}(X)$. In the case of genus 2 Riemann surfaces this is not, however, true. In that case there is an orientation preserving involution $s : \Sigma \to \Sigma$, which is not homotopic to the identity but induces, nevertheless, trivial self–mapping of the Teichmüller space. This reflects the fact that all genus 2 Riemann surfaces are *hyperelliptic*, i.e., they admit an orientation preserving holomorphic involution.

Lemma 4.10.2 *Assume that $[X_n]$ is a converging sequence of points of the Teichmüller space $T(\Sigma)$ of a surface Σ, whose Euler characteristic is negative. Let α be a closed curve on Σ that is not homotopic to a point and let $\ell_\alpha : T(\Sigma) \to \mathbf{R}$ denote the corresponding geodesic length function. Then the sequence $\ell_\alpha(X_n)$ converges to a finite limit.*

Proof. This is obvious and follows from the continuity (Theorem 4.9.2, page 159) of the geodesic length function (which is actually real analytic).

From the definitions it follows immediately that the elements of the modular group $\Gamma^g = \Gamma(\Sigma)$ of a genus g, $g > 1$, surface Σ are homeomorphic self–mappings of the Teichmüller space $T^g = T(\Sigma)$ and isometries of the Teichmüller metric τ. This does not require any proof. The modular group acts on the Teichmüller space.

Lemma 4.10.3 *Let $f_n^* \in \Gamma(\Sigma)$ be elements of the modular group and let $[X] \in T(\Sigma)$ be a point in the Teichmüller space. Then the length spectrum $\mathcal{L}(f_n^*(X))$ of a Riemann surface representing the point $f_n^*([X])$ does not depend on the element $f_n^* \in \Gamma(\Sigma)$.*

Proof. This is also obvious by the definition of the action of the modular group. Recall that an orientation preserving diffeomorphisms $f : \Sigma \to \Sigma$ induces first a mapping (cf. formula (4.8) on page 145) $f^* : \mathcal{M}(\Sigma) \to \mathcal{M}(\Sigma)$ where $f^*(X)$, for a complex structure X, is defined by requiring the mapping $f : (\Sigma, f^*(X)) \to (\Sigma, X)$ be an isometry of the corresponding hyperbolic metrics. This means that the set of lengths of closed geodesic curves on (Σ, X) and on $(\Sigma, f^*(X))$ are equal (as sets).

Lemma 4.10.4 *Assume that $g > 1$. Orbits of points of T^g under the action of the modular group Γ^g are discrete subsets of T^g.*

This is a consequence of the discreteness of the length spectrum (Theorem 3.15.1, page 115).

Assume the contrary. Then one can choose a point $[X] \in T^g$ and a sequence f_n^* of distinct elements of the modular group in such a way that $f_n^*([X])$ converges to a point in T^g and all points $f_n^*([X])$ are *distinct* points of T^g. Let us show that this is not possible.

By Theorem 4.8.2 we can find a set $\{\alpha_1, \ldots, \alpha_{6g-4}\}$ simple closed curves on Σ such that the associated geodesic length functions form an injective mapping of the Teichmüller space into $\mathbf{R}_+^{6g-4}$.

Let $\mathcal{L} = \mathcal{L}(X)$ be the length spectrum of X. By Lemma 4.10.3,

$$\mathcal{L}(f_n^*(X)) = \mathcal{L}$$

for each n. By Theorem 3.15.1, $\mathcal{L}$ is a discrete subset of $\mathbf{R}$.

By Lemma 4.10.2 the sequence $\ell_{\alpha_j}(f_n^*(X))$ converges as $n \to \infty$ for each index j. Since all the values $\ell_{\alpha_j}(f_n^*(X)) \in \mathcal{L}(f_n^*(X)) = \mathcal{L}$ belong to the same discrete subset of $\mathbf{R}$, convergence can happen only if for each j there is an index N_j such that $\ell_{\alpha_j}(f_n^*(X))$ is independent of n for $n > N_j$. Let now $N_0 = \max\{N_1, \ldots, N_{6g-4}\}$. Then, for $n > N_0$, $\ell_{\alpha_j}(f_n^*(X))$ is independent of n for all j. Theorem 4.8.2 implies then that all the points $f_n^*([X]) \in T^g$ agree which is a contradiction.□

Theorem 4.10.5 *Assume that $g > 1$. The action of Γ^g on T^g is properly discontinuous.*

Proof. It clearly suffices to show that each point $[X] \in T^g$ has an open neighborhood U_X such that:

> $f^*(U_X) \cap U_X \neq \emptyset$ for at most finitely many elements f^* of the modular group Γ^g.

To construct such a neighborhood U_X observe first that by Lemma 4.10.4

$$\epsilon_X = \frac{1}{4}\inf\{\tau([X], f^*([X])) \mid f^* \in \Gamma^g,\ f^*([X]) \neq [X]\}$$

is a positive number. Here τ is the Teichmüller metric. Let

$$U_X = \{[Y] \in T^g \mid \tau([X],[Y]) < \epsilon_X\}.$$

Since all elements of the modular group are isometries of the Teichmüller metric, the definition implies that $f^*(\overline{U}_X) \cap \overline{U}_X = \emptyset$ for all $f^* \in \Gamma^g$ not fixing the point $[X]$.

By Lemma 4.10.1 and by Theorem 3.9.3 (page 99), there are at most $84(g-1)$ elements of Γ^g fixing the point $[X]$ proving the theorem.

Here we have considered Teichmüller spaces of compact classical surfaces of genus > 1 simply for technical convenience. It is not hard to see that the same holds for actually all compact surfaces. The case of classical genus 1 surfaces will be dealt with separately. Same arguments can be applied also in the case of non–classical compact surfaces.

Theorem 4.10.5 was first shown by S. Kravetz [54]. The arguments presented here follow the lines of the presentation of F. P. Gardiner [32, Section 8.5].

4.11 Representations of groups

Teichmüller spaces of Riemann surfaces can be studied also in a more abstract setting using representations of groups. That leads to an interesting parametrization of the Teichmüller space of a classical Riemann surface as a *component of an affine real algebraic variety.* This approach is interesting also because it leads to new ways of compactifying the Teichmüller space ([73], [64], [16]). Within this monograph we cannot present these compactifications. We can, however, show that the Teichmüller space is a component of an affine variety. This important result has been shown independently by several authors. To our knowledge the first one to do this was Heinz Helling ([39]). We follow his constructions here.

At this point it is necessary to review results from the theory of deformations of representations of groups. We cannot prove everything here. For more details we refer to [83], [97], [98] and [39]. The following notation and definition is related to Definition 4.7.1 (on page 153) but it is not *exactly* the same.

Definition 4.11.1 *Let* Γ *be a group and* F *a topological group. The* deformation space $\mathcal{R}(\Gamma, F)$ *of* Γ *in* F *consists of all homomorphisms* $\Gamma \to F$.

We endow $\mathcal{R}(\Gamma, F)$ with the topology of point–wise convergence. In this topology a sequence formed of the homomorphisms $\theta_j : \Gamma \to F$, $j = 1, 2, \ldots$, converges to a homomorphism $\theta : \Gamma \to F$ if and only if for each $\gamma \in \Gamma$ the sequence $\theta_1(\gamma), \theta_2(\gamma), \ldots$ converges to $\theta(\gamma)$ in the topological group F.

For a topological group F, $\mathrm{Aut}(F)$ is the group of all continuous automorphisms of F. We denote by $\mathrm{Aut}_0(F)$ the group of inner automorphisms of F.

The group $\mathrm{Aut}(F)$ acts on $\mathcal{R}(\Gamma, F)$ in the natural way: an element $f \in \mathrm{Aut}(F)$ induces the mapping $f^* : \mathcal{R}(\Gamma, F) \to \mathcal{R}(\Gamma, F)$ defined by setting $f^*(\theta) = f \circ \theta$ for every $\theta \in \mathcal{R}(\Gamma, F)$.

Definition 4.11.2 *The quotient*

$$\mathcal{T}(\Gamma, F) = \mathcal{R}(\Gamma, F)/\mathrm{Aut}_0(F)$$

is called the Teichmüller space of representations of Γ in F.

For a locally compact group F, let

$$\mathcal{R}^0(\Gamma, F) = \{\theta \in \mathcal{R}(\Gamma, F) | \theta \text{ injective}, \ \theta(\Gamma) \text{ discrete}, \ F/\theta(\Gamma) \text{ compact}\}.$$

The subspace $\mathcal{R}^0(\Gamma, F)$ is clearly invariant under the action of $\mathrm{Aut}(F)$. We use the notation $\mathcal{T}^0(\Gamma, F)$ for the image of $\mathcal{R}^0(\Gamma, F)$ in $\mathcal{T}(\Gamma, F)$ under the projection $\mathcal{R} \to \mathcal{T}$.

Definition 4.11.3 *This space* $\mathcal{T}^0(\Gamma, F)$ *is called* the Teichmüller space of discrete representations of Γ in F.

In certain special cases the space $\mathcal{T}^0(\Gamma, F)$ is closely related to the usual Teichmüller space of a surface. To see the connection consider the fundamental group Γ (at some implicit base point) of a compact and oriented surface Σ without boundary. Let X be a complex structure on Σ.

By the uniformization theorem we get a presentation $X = U/G$ where U is the upper half–plane and $G \subset \mathrm{PSL}_2(\mathbf{R})$ is a discrete subgroup, i.e., a Fuchsian group. This, in turn, gives rise to an isomorphism (3.6) (see page 87)

$$\theta : \Gamma \to G \subset \mathrm{PSL}_2(\mathbf{R}).$$

If G' is another group for which $X = U/G'$ and $\theta' : \Gamma \to G'$ is an isomorphism then $\theta \circ \theta^{-1} : G \to G'$ is the restriction to G of an inner automorphism of $\mathrm{PSL}_2(\mathbf{R})$.

The isomorphism $\theta : \Gamma \to G$ is characterized by the following property:

Let $[\alpha] \in \Gamma$, α a closed curve, and let $g = \theta([\alpha])$. Any curve $\tilde{\beta}$ joining a point $z \in U$ to $g(z)$ in the upper half plane projects to a curve β on $U/G = X$ which is freely homotopic to the curve α.

It follows that — even thought there is some ambiguity in the choice of the group G and the isomorphism θ — the point $[\theta : \Gamma \to \mathrm{PSL}_2(\mathbf{R})] \in \mathcal{T}^0(\Gamma, \mathrm{PSL}_2(\mathbf{R}))$ depends only on X. A similar argument shows that the point $[\theta : \Gamma \to \mathrm{PSL}_2(\mathbf{R})] \in \mathcal{T}^0(\Gamma, \mathrm{PSL}_2(\mathbf{R}))$ depends only on $[X] \in T(\Sigma)$. We conclude therefore that

$$\rho : T(\Sigma) \to \mathcal{T}^0(\Gamma, \mathrm{PSL}_2(\mathbf{R})),\ [X] \mapsto [\theta : \Gamma \to \mathrm{PSL}_2(\mathbf{R})]$$

is a well defined mapping.

With the help of the uniformization theorem, one can easily show that ρ is injective.

The next thing that we should observe is that

$$\rho(T(\Sigma)) \neq \mathcal{T}^0(\Gamma, \mathrm{PSL}_2(\mathbf{R})).$$

The reason is that when defining the Teichmüller space $T(\Sigma)$ we started with only those complex structures of Σ that agree with the given orientation of Σ. The complex conjugates of such complex structures form a mirror image of $T(\Sigma)$ which can be mapped to $\mathcal{T}^0(\Gamma, \mathrm{PSL}_2(\mathbf{R}))$ as well. It follows that the Teichmüller space $\mathcal{T}^0(\Gamma, \mathrm{PSL}_2(\mathbf{R}))$ has two connected components which are both models for the Teichmüller space $T(\Sigma)$. We denote these components of $\mathcal{T}^0(\Gamma, \mathrm{PSL}_2(\mathbf{R}))$ by $\mathcal{T}^0_+(\Gamma, \mathrm{PSL}_2(\mathbf{R}))$ and $\mathcal{T}^0_-(\Gamma, \mathrm{PSL}_2(\mathbf{R}))$. For more details we refer to [25].

For our purposes it is better to study representations in $\mathrm{SL}(2, \mathbf{R})$ instead of the representations in $\mathrm{PSL}(2, \mathbf{R})$. By Theorem 3.22.1 (on page 136), in the case of the fundamental group Γ of an oriented compact surface without boundary, every (faithful) representation $\theta : \Gamma \to \mathrm{PSL}_2(\mathbf{R})$ can be lifted to a (faithful) representation $\tilde{\theta} : \Gamma \to \mathrm{SL}_2(\mathbf{R})$. On the other hand, every (faithful) representation $\tilde{\theta} : \Gamma \to \mathrm{SL}_2(\mathbf{R})$ projects to a (faithful) representation $\theta : \Gamma \to \mathrm{PSL}_2(\mathbf{R})$ because the center of the fundamental group of a compact and oriented surface of genus > 1 is trivial.

It follows, therefore, that the projection

$$\mathcal{R}(\Gamma, \mathrm{SL}(2, \mathbf{R})) \to \mathcal{R}(\Gamma, \mathrm{PSL}(2, \mathbf{R}))$$

is surjective. This projection is, of course, continuous and open. It is also obvious that $\mathcal{R}^0(\Gamma, \mathrm{SL}(2, \mathbf{R}))$ projects to $\mathcal{R}^0(\Gamma, \mathrm{PSL}(2, \mathbf{R}))$.

Since

$$\mathcal{T}^0(\Gamma, \mathrm{PSL}(2, \mathbf{R}))$$

has two connected components,

$$\mathcal{T}^0(\Gamma, \mathrm{SL}(2, \mathbf{R}))$$

has only finitely many components. The part of $\mathcal{T}^0(\Gamma, \mathrm{SL}(2,\mathbf{R}))$ projecting to $\mathcal{T}_+^0(\Gamma, \mathrm{PSL}(2,\mathbf{R}))$ will be denoted by $\mathcal{T}_+^0(\Gamma, \mathrm{SL}(2,\mathbf{R}))$. Since $\mathrm{SL}(2,\mathbf{R})$ is connected, the projection

$$\mathcal{R}^0(\Gamma, \mathrm{SL}(2,\mathbf{R})) \to \mathcal{T}^0(\Gamma, \mathrm{SL}(2,\mathbf{R}))$$

defines a bijective correspondence between the components of

$$\mathcal{R}^0(\Gamma, \mathrm{SL}(2,\mathbf{R}))$$

and those of $\mathcal{T}^0(\Gamma, \mathrm{SL}(2,\mathbf{R}))$.

Let G_n be the group generated by the rotation of the upper half-plane around the point i by the rational angle m_n/n in the positive direction. Assume that m_n and n are relatively prime integers and *that the sequence m_n/n converges to an irrational number s.* Then the generators g_n of the groups G_n form a converging sequence and $\lim_{n\to\infty} g_n = g_s$, which is a rotation by the angle s. So, in some sense, the discrete groups G_n 'converge' to the group $\langle g_s \rangle$ which is not discrete.

This is, nevertheless, possible only if we allow G_n *change as a group.* Above group G_n is a cyclic group of order n. For different values of n these groups are not isomorphic to each other.

In the case we are considering the situation is different. The following has been shown in [39].

Theorem 4.11.1 *Connected components of $\mathcal{R}^0(\Gamma, \mathrm{SL}(2,\mathbf{R}))$ are connected components of $\mathcal{R}(\Gamma, \mathrm{SL}(2,\mathbf{R}))$.*

Proof. We have to show that discrete faithful represtations of Γ in $\mathrm{SL}_2(\mathbf{R})$ form an open and closed set in the space of all represtations.

The fact that discrete representations form an open set is an important result of A. Weil who showed in [97, §1] that

$$\mathcal{R}^0(\Gamma, \mathrm{SL}(2,\mathbf{R}))$$

is open in $\mathcal{R}(\Gamma, \mathrm{SL}(2,\mathbf{R}))$. We will not reproduce his proof here.

It remains to show that

$$\mathcal{R}^0(\Gamma, \mathrm{SL}(2,\mathbf{R}))$$

is also closed in $\mathcal{R}(\Gamma, \mathrm{SL}(2,\mathbf{R}))$. In the case that we are considering, namely that of a fundamental group Γ of a compact surface, this fact follows immediately from Theorem 3.11.2 (on page 106).

Assume that $\theta_n : \Gamma \to \mathrm{SL}_2(\mathbf{R})$, $n = 1, 2, \ldots$, is a sequence of faithful representations such that for each n, $U/\theta_n(\Gamma)$ is a compact Riemann surface. Then by Theorem 3.12.1, all non-identity elements of each $\theta_n(\Gamma)$ are hyperbolic Möbius transformations. If $\theta_n \to \theta$ as $n \to \infty$, then, by the definition of the topology, $\theta_n(\gamma) \to \theta(\gamma)$ for each $\gamma \in \Gamma$. The matrices $\theta_n(\gamma)$, $\gamma \in \Gamma \setminus \{\mathrm{Id}\}$, correspond to hyperbolic Möbius transformations. A sequence of hyperbolic Möbius transformations can converge only to one of the following Möbius transformations:

- a hyperbolic Möbius transformation,
- a parabolic Möbius transformation,
- the identity.

It follows, especially, that the group $\theta(\Gamma)$ cannot contain matrices corresponding to elliptic Möbius transformations. Therefore it follows, by Theorem 3.11.2, that the Möbius group corresponding to $\theta(\Gamma)$ is discrete.

The projection $\mathcal{R}(\Gamma, \mathrm{SL}(2,\mathbf{R})) \to \mathcal{T}(\Gamma, \mathrm{SL}(2,\mathbf{R}))$ is open. Therefore the components of $\mathcal{T}^0(\Gamma, \mathrm{SL}(2,\mathbf{R}))$ and those of $\mathcal{T}^0_+(\Gamma, \mathrm{SL}(2,\mathbf{R}))$ are also components of $\mathcal{T}(\Gamma, \mathrm{SL}(2,\mathbf{R}))$.

The following lemma characterizes components of $\mathcal{T}^0_+(\Gamma, \mathrm{SL}(2,\mathbf{R}))$:

Lemma 4.11.2 *Assume that the points corresponding to the representations $\theta : \Gamma \to \mathrm{SL}_2(\mathbf{R})$ and $\theta' : \Gamma \to \mathrm{SL}_2(\mathbf{R})$ belong to the same component of $\mathcal{T}^0_+(\Gamma, \mathrm{SL}(2,\mathbf{R}))$. Then, for any $[\alpha] \in \Gamma$, traces of the matrices $\theta([\alpha])$ and $\theta'([\alpha]) \in \mathrm{SL}_2(\mathbf{R})$ have the same sign.*

Proof. Assume the contrary. Then there exists an $\alpha \in \Gamma$ and representations $\theta : \Gamma \to \mathrm{SL}_2(\mathbf{R})$ and $\theta' : \Gamma \to \mathrm{SL}_2(\mathbf{R})$ which belong to the same component of $\mathcal{T}^0_+(\Gamma, \mathrm{SL}(2,\mathbf{R}))$ and for which the traces of the matrices $\theta(\alpha)$ and $\theta'(\alpha)$ have opposite signs.

Since θ and θ' belong to the same component of $\mathcal{T}^0_+(\Gamma, \mathrm{SL}(2,\mathbf{R}))$ we have a continuous mapping

$$[0,1] \to \mathcal{T}^0_+(\Gamma, \mathrm{SL}(2,\mathbf{R})),\, t \mapsto (\theta_t : \Gamma \to \mathrm{SL}_2(\mathbf{R}))$$

such that $\theta_0 = \theta$ and $\theta_1 = \theta'$.

Then the mapping

$$[0,1] \to \mathbf{R},\ t \mapsto \chi(\theta_t(\alpha))$$

is continuous as well.

Here $\chi(\theta_t(\alpha))$ is the trace of the matrix $\theta_t(\alpha) \in \mathrm{SL}_2(\mathbf{R})$. Because of the assumptions concerning α, θ and θ' this mapping changes sign on $[0,1]$. Therefore we can find an $s \in [0,1]$ such that $\chi(\theta_s(\alpha)) = 0$.

Since, for every $t \in [0,1]$, we have

$$\theta_t \in \mathcal{T}_+^0(\Gamma, \mathrm{SL}(2,\mathbf{R})),$$

every matrix $\theta_t(\alpha)$, $\alpha \in \Gamma$, corresponds to a Möbius transformation in the covering group of a compact and oriented surface of genus > 1. Every such Möbius transformation is hyperbolic, i.e., the trace of such a matrix has to have absolute value ≥ 2. We have therefore reached a contradiction proving the lemma.

Let us now take a closer look at the projection $\mathcal{T}^0(\Gamma, \mathrm{SL}(2,\mathbf{R})) \to \mathcal{T}^0(\Gamma, \mathrm{PSL}(2,\mathbf{R}))$. We make first the following observation: *Representations $\theta_j : \Gamma \to \mathrm{SL}(2,\mathbf{R})$, $j = 1, 2$, project to the same representation $\Gamma \to \mathrm{PSL}(2,\mathbf{R})$ if and only if there exists a function ξ on Γ, taking the values ± 1, such that $\theta_1(\gamma) = \xi(\gamma)\theta_2(\gamma)$ for all $\gamma \in \Gamma$.* Using this remark and the reasoning of the above lemma we prove:

Theorem 4.11.3 *Let Σ be a compact and oriented surface of genus > 1. Assume that $\Gamma \subset \mathrm{SL}(2,\mathbf{R})$ is isomorphic to the fundamental group of the surface Σ with some base point. The space $\mathcal{T}_+^0(\Gamma, \mathrm{SL}(2,\mathbf{R}))$ has finitely many components each of which is homeomorphic to the Teichmüller space of Σ.*

Proof. The natural projection

$$\Pi : \mathcal{T}_+^0(\Gamma, \mathrm{SL}(2,\mathbf{R})) \to \mathcal{T}_+^0(\Gamma, \mathrm{PSL}(2,\mathbf{R}))$$

is continuous and open. It is clear that $\mathcal{T}_+^0(\Gamma, \mathrm{SL}(2,\mathbf{R}))$ has only finitely many components. It suffices to show that no components of

$$\mathcal{T}_+^0(\Gamma, \mathrm{SL}(2,\mathbf{R}))$$

contain two different points which project onto the same point of

$$\mathcal{T}_+^0(\Gamma, \mathrm{PSL}(2,\mathbf{R})).$$

To that end let $\overline{\theta} : \Gamma \to \mathrm{PSL}(2,\mathbf{R})$ represent a point of

$$\mathcal{T}_+^0(\Gamma, \mathrm{PSL}(2,\mathbf{R})).$$

Let $\theta : \Gamma \to \mathrm{SL}(2,\mathbf{R})$ be a representation that projects to $\overline{\theta}$. If $\theta' : \Gamma \to \mathrm{SL}(2,\mathbf{R})$ is another representation projecting also to $\overline{\theta}$ then there exists a function ξ on Γ taking the values ± 1 such that $\theta'(\gamma) = \xi(\gamma)\theta(\gamma)$ holds for all $\gamma \in \Gamma$. Let us show that there is no continuous path, in $\mathcal{T}_+^0(\Gamma, \mathrm{SL}(2,\mathbf{R}))$, joining two different points of $\mathcal{T}_+^0(\Gamma, \mathrm{SL}(2,\mathbf{R}))$, which both project to the same point in $\mathcal{T}_+^0(\Gamma, \mathrm{PSL}(2,\mathbf{R}))$. Assume that one these points corresponds to θ and the other to θ'.

To that end, consider the function ξ satisfying $\theta' = \xi \cdot \theta$. Since $\theta \neq \theta'$ there exists an $\gamma \in \Gamma$ such that $\xi(\gamma) = -1$. Considering the function

$$t_\gamma : \mathcal{T}_+^0(\Gamma, \mathrm{SL}(2,\mathbf{R})) \to \mathbf{R}, [\theta] \mapsto \chi\,\theta(\gamma)$$

and repeating the argument of Lemma 2.1 we conclude that $[\theta]$ and $[\theta']$ do not belong to the same component of $\mathcal{T}_+^0(\Gamma, \mathrm{SL}_2(\mathbf{R}))$.

4.12 The algebraic structure

The Teichmüller space of a compact and oriented surface Σ can be given the local structure of an affine real algebraic variety. We will use here traces of elements of $\mathrm{SL}_2(\mathbf{R})$ to parametrize the Teichmüller space and to embed it into an affine space $\mathbf{R}^M$ in such a way that it becomes a component of an affine variety. The construction that we review here is due to Heinz Helling ([39]).

Let Γ be again the fundamental group of a compact and oriented surface Σ. Let $\alpha_1, \ldots, \alpha_m$ be any set of generators for Γ satisfying certain defining relations.

Let I_m denote the ordered set of all ordered j-tupels

$$J^{\#} = (\nu_1, \nu_2, \ldots, \nu_j)$$

of natural numbers with $1 \leq \nu_1 < \nu_2 < \cdots < \nu_j \leq m$, $1 \leq j \leq m$. Let $K = K(m)$ be the set

$$K = \{\alpha_{\nu_1}\alpha_{\nu_2}\ldots\alpha_{\nu_j} | (\nu_1, \nu_2, \ldots, \nu_j) \in I_m\}.$$

For any representation $\theta : \Gamma \to \mathrm{SL}_2(\mathbf{R})$ we may form the function

$$\Gamma \to \mathbf{R},\ \gamma \mapsto \chi(\theta(\gamma)).$$

(Here $\chi(\theta(\gamma))$ is the trace of the matrix $\theta(\gamma) \in \mathrm{SL}_2(\mathbf{R})$.) This function is referred to as a *trace function.*

Define the mapping

$$\mathcal{R}(\Gamma, \mathrm{SL}(2,\mathbf{R})) \to \mathbf{R}^{2^m-1}, \theta \mapsto (\ldots, \chi(\theta(\alpha_{\nu_1}\alpha_{\nu_2}\ldots\alpha_{\nu_j})), \ldots) \in \mathbf{R}^{2^m-1} \tag{4.21}$$

$$(\nu_1, \nu_2, \ldots, \nu_j) \in I_m.$$

Since traces are invariant under conjugation, this defines a mapping

$$h_K : \mathcal{T}(\Gamma, \mathrm{SL}(2, \mathbf{R})) \to \mathbf{R}^{2^m-1}.$$

Let A_K be the image of $\mathcal{T}(\Gamma, \mathrm{SL}(2, \mathbf{R}))$ under this mapping, and let $\mathcal{J}$ be the ideal of polynomials vanishing on A_K. These polynomials consist of all relations between the values of traces functions on Γ. By the properties of traces, these relations are polynomials with rational coefficients. For a detailed discussion of these facts see Section B.1 in Appendix B.[2]

Let V_Γ be the set of zeros of $\mathcal{J}$ in $\mathbf{R}^{2^m-1}$. Clearly $\mathcal{R}(\Gamma, \mathrm{SL}(2, \mathbf{R}))$ maps into V_Γ under the mapping (4.21), i.e.,

$$h_K(\mathcal{T}(\Gamma, \mathrm{SL}(2, \mathbf{R})) \subset V_\Gamma.$$

The above inclusion is actually an equality, i.e.,

$$h_K(\mathcal{T}(\Gamma, \mathrm{SL}(2, \mathbf{R})) = V_\Gamma.$$

This is not difficult to see and has been shown in Appendix B, Corollary B.1.2 on page 247.

Let

$$V_\Gamma^0 = h_K\left(\mathcal{T}_+^0(\Gamma, \mathrm{SL}(2, \mathbf{R}))\right).$$

In view of Theorem 4.11.1 we have:

Theorem 4.12.1 *The set V_Γ^0 is a union of components of the affine real algebraic variety V_Γ. The mapping*

$$h_K : \mathcal{T}_+^0(\Gamma, \mathrm{SL}(2, \mathbf{R})) \to V_\Gamma^0$$

is a homeomorphism with respect to the Hausdorff topology of the affine space containing V_Γ.

By Theorem 4.11.3 we have the following:

Theorem 4.12.2 *Each component of V_Γ^0 is homeomorphic to the Teichmüller space $T(\Sigma)$.*

In this way the Teichmüller space becomes a component of an affine real algebraic variety. The construction involves some choices like the choice of generators for Γ. Heinz Helling ([39]) has shown that this construction is essentially independent of these choices. And that the action of the modular group on this algebraic model for $T(\Sigma)$ is biregular and can be defined by polynomials with rational coefficients.

[2] In [40] Helling has even computed these polynomials in the case of genus 2 surfaces.

4.13 Reduction of parameters

In this section we continue to study oriented compact surfaces without boundary. In the above construction we obtained, via the mapping h_K, a presentation of the Teichmüller space of such an oriented surface Σ of genus g (and without boundary) as a component of an affine real algebraic variety in an affine space $\mathbf{R}^M$ for a rather large M. In this chapter we show that the embedding of the Teichmüller space can be chosen in such a way that the codimension of the image of the Teichmüller space will be two.

It is our conjecture that, in the case of oriented compact surfaces without boundary this is also the smallest possible codimension that can occur when parametrizing Teichmüller spaces of such surfaces by trace functions.

Let X be a complex structure on a compact and oriented surface Σ which does not have any boundary components and whose genus g is > 1. For any element $[\alpha] \in \Gamma$ of the fundamental group of Σ let $\ell_\alpha(X)$ denote, as before, the length of the geodesic curve homotopic to α on X. Theorem 1.6.4 (on page 46) gives a set $K' = \{\alpha_1, \ldots, \alpha_{6g-4}\}$ of closed curves representing elements of Γ such that the associated mapping

$$\mathcal{L}_{K'} : T(\Sigma) \to \mathbf{R}^{6g-4}, \ [X] \mapsto (\ell_{\alpha_1}(X), \ldots, \ell_{\alpha_{6g-4}}(X))$$

is injective.

The various mappings of the above construction can be collected to the following diagram:

$$\begin{array}{ccccccc}
\mathcal{M}(\Sigma) & \rightarrow & \mathcal{R}^0(\Gamma, \mathrm{SL}(2,\mathbf{R})) & \subset & \mathcal{R}(\Gamma, \mathrm{SL}(2,\mathbf{R})) & & \\
 & & \downarrow & & \downarrow & & \\
\downarrow & & \mathcal{T}^0(\Gamma, \mathrm{SL}_2(\mathbf{R})) & \subset & \mathcal{T}(\Gamma, \mathrm{SL}_2(\mathbf{R})) & \xrightarrow{h_K} & \mathbf{R}^{2^m-1} \\
 & \nearrow & & & & & \\
T(\Sigma) & & & & & &
\end{array}$$

On the other hand we may form the mapping $h_{K'} : \mathcal{T}(\Gamma, \mathrm{SL}_2(\mathbf{R})) \to \mathbf{R}^{6g-4}$ replacing K by K' in the construction of h_K.

We consider this mapping and show first the following result:

Lemma 4.13.1 *The mapping $h_{K'}$ is injective on $\mathcal{T}^0(\Gamma, \mathrm{SL}_2(\mathbf{R}))$.*

Proof. We first have to observe that the set K' certainly contains generators for Γ. By Lemma 2.1, for each j the traces of the matrices $\theta([\alpha_j])$, $[\alpha_j] \in K'$, are either positive for all $[\theta]$ in a fixed component of $\mathcal{T}^0(\Gamma, \mathrm{SL}_2(\mathbf{R}))$ or all negative. Signs of the traces of the matrices $\theta([\alpha_j])$ are determined by the component of $\mathcal{T}^0(\Gamma, \mathrm{SL}_2(\mathbf{R}))$ to which $[\theta]$ belongs. These signs also determine the corresponding component. It follows therefore that the images, under the mapping $h_{K'} : \mathcal{T}^0(\Gamma, \mathrm{SL}_2(\mathbf{R})) \to \mathbf{R}^{6g-4}$, of distinct components of $\mathcal{T}^0(\Gamma, \mathrm{SL}_2(\mathbf{R}))$ are distinct.

It suffices to show therefore that the mapping $h_{K'}$ is injective when restricted to a component of $\mathcal{T}^0(\Gamma, \mathrm{SL}_2(\mathbf{R}))$.

Let now $[\theta] \in \mathcal{T}^0(\Gamma, \mathrm{SL}_2(\mathbf{R}))$ be a representation corresponding to the point $[X]$ in $T(\Sigma)$. Then,

$$\ell_\alpha(X) = \log\left(\frac{1}{2}\left(|\mathrm{tr}\,\theta(\alpha)| + \sqrt{(\mathrm{tr}\,\theta(\alpha))^2 - 4}\right)\right).$$

Define the mapping $W : \mathbf{R}^{6g-4} \to \mathbf{R}^{6g-4}$ setting

$$W(\ldots, r_j, \ldots) = \left(\ldots, \log\left(\frac{1}{2}\left(|r_j| + \sqrt{r_j^2 - 4}\right)\right), \ldots\right). \tag{4.22}$$

Then we have

$$\mathcal{L}_{K'} = W \circ h_{K'}.$$

Since $\mathcal{L}_{K'}$ is injective by Theorem 1.6.4 (on page 46), also $h_{K'}$ has to be injective on a component of $\mathcal{T}(\Gamma, \mathrm{SL}_2(\mathbf{R}))$. This proves the lemma.

Let $\mathbf{R}^k \subset \mathbf{R}^{2^m-1}$ be the corresponding subspace and let $pr : \mathbf{R}^{2^m-1} \to \mathbf{R}^k$ be the projection onto this subspace. The above statement simply means that the mapping

$$h_{K'} = pr \circ h_K : T(\Sigma) \to \mathbf{R}^k$$

is injective.

The image of an affine variety under a projection is, on the other hand, still an affine variety. Therefore $V_{K'} = pr(V_\Gamma) \subset \mathbf{R}^k$ is an affine variety. It is clear by the construction that $pr \circ h_K(T(\Sigma))$ is contained in a component of this variety $V_{K'}$.

The projection being an open mapping, $pr \circ h_K(T(\Sigma))$ is an open subset of V_K. Repeating the arguments of the proof of Theorem 4.11.1, it follows finally that $pr \circ h_K(T(\Sigma))$ is also closed in V_K.

Therefore we have:

Theorem 4.13.2 *The image of the Teichmüller space $T(\Sigma)$ under the mapping $h_{K'}$ is a component of an affine real algebraic variety of codimension 2.*

Proof. The only thing that remains to be shown is the statement concerning the codimension. But that is immediate since the dimension of the Teichmüller space is $6g - 6$ while the dimension of the ambient affine space is $6g - 4$. The theorem is therefore proved.

It is also obvious that all the results of Heinz Helling hold also for this presentation of the Teichmüller space. In particular, the modular group acts as a group of biregular self mappings of $V_{K'}$ and maps the part of $V_{K'}$ corresponding to the Teichmüller space $T(\Sigma)$ onto itself. This is actually quite straightforward, Helling's arguments can be repeated here word by word ([39]).

4.14 Extension to non–classical surfaces

In this Section we show that even the Teichmüller spaces of non–classical surfaces can be given an affine structure in a natural way. We cannot, however, proceed here by simply repeating the preceding arguments for non–classical surfaces. The above construction was based on a result of A. Weil which does not hold for surfaces with boundary. Using Theorem 4.5.1 (page 149) we can deal with this difficulty.

So let Σ be a compact non–classical surface of genus g. Let Σ^c be its complex double and $\sigma : \Sigma^c \to \Sigma^c$ the orientation reversing involution for which $\Sigma = \Sigma^c/\langle\sigma\rangle$.

Consider the presentation for the Teichmüller space $T(\Sigma^c)$ as a component of an affine variety. It depends on various choices like the choice of the set K of generators for the fundamental group of Σ^c. The affine structure is, nevertheless, independent of this choice. So we may start with any generating set K. In particular we may choose K in such a way that the involution σ maps the elements of K onto elements of K. Therefore σ defines a permutation of the elements of K.

Checking through the construction it is now obvious that we have the following commutative diagram:

$$\begin{array}{ccccc} T(\Sigma^c) & \longrightarrow & V_\Gamma & \subset & \mathbf{R}^M \\ \downarrow{\scriptstyle \sigma^*} & & \downarrow{\scriptstyle \text{per}} & & \downarrow{\scriptstyle \text{per}} \\ T(\Sigma^c) & \longrightarrow & V_\Gamma & \subset & \mathbf{R}^M \end{array}$$

where per : $\mathbf{R}^M \to \mathbf{R}^M$ is a permutation of coordinates of $\mathbf{R}^M$, Γ is the fundamental group of Σ^c and V_Γ is a real affine variety.

By Theorem 4.12.2 the image of $T(\Sigma^c)$ in V_Γ is a component of V_Γ.

On the other hand, by Theorem 4.5.1, the Teichmüller space $T(\Sigma)$ of the surface Σ can be identified with the fixed–point set $T(\Sigma^c)_{\sigma^*}$ of the involution $\sigma^* : T(\Sigma^c) \to T(\Sigma^c)$.

Now the fixed–point set of a permutation of coordinates per : $V_\Gamma \to V_\Gamma$ is trivially an affine variety, denote that by $(V_\Gamma)_{\text{per}}$. Since the above diagram commutes we conclude finally from Theorem 4.12.2 that $T(\Sigma^c)_{\sigma^*}$ is a component of the affine real algebraic variety $(V_\Gamma)_{\text{per}}$.

This affine structure depends, a priori, on all the choices made. In view of the results of Helling it follows, however, that the various affine structures obtained for $T(\Sigma^c)_{\sigma^*}$ in this way are isomorphic. We have therefore:

Theorem 4.14.1 *The Teichmüller space of a non-classical compact surface can be represented as a component of an affine real algebraic variety.*

Chapter 5

Moduli spaces

5.1 Introduction to Chapter 5

Points of the Teichmüller T^g space of a genus g surface Σ are isomorphism classes of *marked* complex structures of the surface Σ. Here two complex structures X and Y define the same point of the Teichmüller space if there is a holomorphic homeomorphism $f : (\Sigma, X) \to (\Sigma, Y)$, *which is homotopic to the identity.* This additional topological condition "homotopic to the identity" was enforced in order to get a nice structure for the parametrizing space as was explained in the introduction to Chapter 4. As we have seen in that Chapter, the Teichmüller space of a compact genus g surface is a complex manifold (of complex dimension $3g - 3$) homeomorphic to an euclidean space.

From the point of view of Riemann surfaces, this topological condition is, however, not natural. The natural object of study is the moduli space, M^g of genus g Riemann surfaces, i.e., the space of isomorphism classes of complex structures of a genus g surface. This space is, however, harder to study than the Teichmüller space.

In this Chapter we will consider the structure of the moduli space of compact Riemann surfaces. We show that the moduli space of compact classical genus g surfaces is a connected but not compact Hausdorff space. In the same way on can also study the moduli space of compact non–classical genus g surfaces. It turns out that this moduli space is a non–compact Hausdorff space with $\lfloor (3g + 4)/2 \rfloor$ connected components.

After these observations, which follow rather quickly from Teichmüller theory, we turn our attention to the question of finding a geometrically natural compactifications for these moduli spaces. We construct them using straightforward geometric methods. It turns out that both moduli spaces, that of classical genus g surfaces and that of non–classical genus g surfaces, admit a similar compactification which makes them both connected and compact Hausdorff spaces. This construction relies on considerations of Chapter 3, especially on Theorem 3.19.1 (on page 131), which plays a key rôle here.

We conclude this Chapter by considering the analytic structure of the moduli space of classical and non–classical genus g Riemann surfaces.

5.2 Moduli spaces of smooth Riemann surfaces

In this section we suppose everywhere that the surfaces that we are considering are compact and have negative Euler characteristic.

Sometimes, when we want to stress the fact that we are dealing with ordinary Riemann surfaces, we may use the adjective 'smooth' as contrary to 'non–smooth' Riemann surfaces which are allowed to have simple singularities. In this Section we study only these ordinary smooth Riemann surfaces.

Recall the definitions for the Teichmüller space (Definition 4.3.1, on page 145) the moduli space (Definition 4.3.3) and the modular group of a compact surface Σ. By the definitions it is immediate that

$$M(\Sigma) = T(\Sigma)/\Gamma(\Sigma). \tag{5.1}$$

Let us first consider a classical compact genus g surface Σ. For such a surface Σ we use the notations $\Gamma^g = \Gamma(\Sigma)$, $T^g = T(\Sigma)$ and $M^g = M(\Sigma)$. By (5.1) we can derive results concerning the moduli space by using those of the Teichmüller space. Recall first that by Theorem 4.10.5 on page 163, Γ^g acts properly discontinuously on the Teichmüller space T^g, which in this case is a is a complex manifold of complex dimension $3g - 3$ (for a description of the complex structure see page 148).

Using (5.1) we equip the moduli space with the quotient topology. Elements of the modular group are isometries of the Teichmüller metric. Therefore the Teichmüller metric of T^g induces also a metric on M^g, which is, therefore, a metric space. We have:

Lemma 5.2.1 *The moduli space M^g of a compact and classical genus g surface is a connected and non–compact metric space.*

Proof. By the above remarks, the moduli space $M^g = T^g/\Gamma^g$ is a metric space. It is connected, since the Teichmüller space T^g is connected by Theorem 4.4.1 (on page 147).

Thus the only thing that remains to be shown is the statement about the non–compactness of the moduli space. To that end, consider first the

Teichmüller space T^g. Define the function

$$\epsilon : T^g \to \mathbf{R}_+,\ [X] \mapsto \min\{\ell_\alpha(X) \mid \alpha \text{ a closed geodesic curve on } X\}$$

For a Riemann surface X, $\epsilon([X])$ is the length of the shortest simple closed geodesic curve on X. Recall that by Corollary 3.15.2 (on page 116) there are always curves with minimal length on any compact hyperbolic Riemann surface.

Using the discreteness of the length spectrum and the continuity of the geodesic length functions one can easily show that locally the function ϵ is the minimum of a finite number of geodesic length functions. The function ϵ is, therefore, continuous. It is also equivariant with respect to the action of the modular group, i.e.,

$$\epsilon \circ f^* = \epsilon,\ \forall f^* \in \Gamma^g.$$

We conclude that ϵ defines a continuous function $M^g \to \mathbf{R}_+$. We use the notation ϵ also for this function.

Assume that M^g is compact. Then the continuous function ϵ attains its minimum ϵ_{min} at some point $[X] \in M^g$. This minimum is, of course, a positive number. This would imply that all non–trivial simple closed curves on any compact Riemann surface of genus g have length $\geq \epsilon_{min}$. But this is absurd. Using pairs of pants with short boundary geodesics it is easy to build Riemann surfaces of any genus g, $g > 1$, such that these surfaces have arbitrarily short non–trivial simple closed geodesics.

The above proof of Lemma 5.2.1 suggests that an ideal boundary of the moduli space of genus g Riemann surfaces, $g > 1$, consists of points corresponding to such singular Riemann surfaces where we have replaced certain simple closed geodesic curves by points, i.e., the length of these geodesics $\to 0$ as we approach this boundary point of the moduli space. That this is, indeed, the case will be shown in the subsequent sections. Main problem here is to get a good geometric picture about all the possible degenerations of compact Riemann surfaces. Considerations of Chapter 3 play an important rôle here.

The constructions that we will present for the moduli spaces of classical Riemann surfaces can be generalized to the case of the moduli spaces of non–classical Riemann surfaces of a given genus g. Most important problems here are related with the compactification. Before taking up these matters we will consider the structure of the moduli spaces of smooth non–classical Riemann surfaces.

5.3 Moduli spaces of genus 1 surfaces

Moduli spaces of genus 1 Riemann surfaces are quite well understood. In order to motivate the subsequent considerations we will, in this chapter, review this, more classical, part of the moduli problems.

Recall that genus 1 Riemann surfaces are:

- torus,
- annulus,
- Klein bottle,
- Möbius band.

Moduli spaces of the non–classical genus 1 surfaces, the annulus, the Klein bottle and the Möbius band, will be studied by embedding them into the moduli space of the torus. This embedding will be constructed by means of the complex double and is based on Theorem 4.5.1 (on page 149). Therefore we start with considering the moduli space of the torus.

Let T^1 be a topological torus. Fix a basis $\langle \alpha_1, \alpha_2 \rangle$ for the fundamental group of T^1. Here we use the notation α_j for a closed curve α_j and for its homotopy class. Let X a complex structure on T^1. The Riemann surface $X = (T^1, X)$ has the finite complex plane $\mathbf{C}$ as its universal covering surface and

$$X = \mathbf{C}/L_\tau,$$

where L_τ is a lattice generated by two translations $g(z) = z + \omega_1$ and $h(z) = z + \omega_2$. We may, furthermore, assume that $z \mapsto z + \omega_j$ covers the homotopy class of α_j.

The quotient $\mathbf{C}/L_\tau$ is a torus if and only if $\tau := \omega_1/\omega_2$ is not real. Without restricting the generality we may also assume that $\omega_1 = 1$ and $\operatorname{Im} \tau > 0$. Then $L_\tau = \mathbf{Z} + \mathbf{Z} \cdot \tau$ and the mapping

$$T^1 \to U, \;\; [\mathbf{C}/L_\tau] \mapsto \tau$$

is a bijection and the Teichmüller space of the torus can be, via this mapping, identified with the upper half–plane U (Lemma 4.6.1, page 152).

We want ot get our hands on the moduli space of the torus. To that end we need to get a sufficient and necessary condition which guarantees that two given torii are isomorphic. That is done in the next lemma. Suppose, that $\operatorname{Im} \tau > 0$. Let $L_\tau = \mathbf{Z} + \mathbf{Z} \cdot \tau$ and $L_{\tau'} = \mathbf{Z} + \mathbf{Z} \cdot \tau'$.

Lemma 5.3.1 *There is a holomorphic homeomorphism between the torii $T = \mathbf{C}/L_\tau$ and $T = \mathbf{C}/L_{\tau'}$ if and only if there exists a matrix*

$$\begin{pmatrix} a & b \\ c & d \end{pmatrix} \in \mathrm{SL}_2(\mathbf{Z})$$

such that

$$\tau' = \frac{a\tau + b}{c\tau + d}. \tag{5.2}$$

Proof. Straightforward computation.

Definition 5.3.1 *The group* $\mathrm{SL}_2(\mathbf{Z})$ *whose elements are Möbius transformations*

$$z \mapsto \frac{az + b}{cz + d}$$

where a, b, $c\,d \in \mathbf{Z}$ and $ad - bc = 1$ is called the elliptic modular group.

The elliptic modular group acts on U and is discrete. It is properly discontinuous by Theorem 3.5.3 (page 83).

For a lattice $L_\tau = \mathbf{Z} + \mathbf{Z} \cdot \tau$ consider the numbers

$$g_2 = 60 \sum_{\substack{\lambda \in L \\ \lambda \neq 0}} \frac{1}{\lambda^4}$$

$$g_3 = 140 \sum_{\substack{\lambda \in L \\ \lambda \neq 0}} \frac{1}{\lambda^6}.$$

Definition 5.3.2 *Let*

$$j(X) = j(\tau) = \frac{1728 \cdot g_2^3}{g_2^3 - 27 g_3^2}$$

for a torus $X = \mathbf{C}/L_\tau$, $L_\tau = \mathbf{Z} + \mathbf{Z} \cdot \tau$.

The function $j(\omega)$ is the elliptic modular function.

Theorem 5.3.2 *Torii*

$$\mathbf{C}/(\mathbf{Z} + \mathbf{Z} \cdot \tau) \text{ and } \mathbf{C}/(\mathbf{Z} + \mathbf{Z} \cdot \tau')$$

are isomorphic if and only if $j(\tau) = j(\tau')$.

For a proof we refer to the monograph of Siegel [86, Theorem 1 on page 79] and that of Alling [7, Theorem 9.29, page 181]. The latter contains a detailed study of the elliptic modular function.

From Lemma 5.3.1 and from Theorem 5.3.2 it follows that j defines a mapping $j : M^1 \to \mathbf{C}$ which is actually a bijection. We have, furthermore, the commutative diagram:

$$\begin{array}{ccc} U & & \\ \downarrow \pi & \searrow \tilde{j} & \\ M^1 & \xrightarrow{j} & \mathbf{C} \end{array} \tag{5.3}$$

This is the classical construction for the moduli space of tori. The mapping $\tilde{j} : U \to \mathbf{C}$ is a smooth covering map whose cover group is the elliptic modular group. This discontinuous group contains the parabolic Möbius transformation $z \mapsto z + 1$ as primitive element. By Lemma 3.10.2, a fundamental domain for this group contains the domain $\{z \in \mathbf{C} \mid \operatorname{Im} z > 1, \ |\operatorname{Re} z| < \frac{1}{2}\}$. This group serves also to show that the half–plane H in Lemma 3.10.2 is the largest possible, since a closer analysis shows that a fundamental domain for the elliptic modular group is the interior of

$$W = \{\omega \in U | -1/2 < \operatorname{Re}\omega \leq 1/2, |\omega| \geq 1 \text{ and } |\omega| > 1 \text{ for } \operatorname{Re}\omega \leq 0\}.$$

A detailed proof for this fact can be found in the monograph of Siegel [87, Theorem 3, section 9] or in that of Norman Alling [7, Chapter 9].

We have observed above that the elliptic modular function furnishes a bijection $j : M^1 \to \mathbf{C}$. The complex plane can, on the other hand, be compactified to the Riemann sphere $\hat{\mathbf{C}}$ by adding the point at the infinity. That is also the usual compactification for M^1.

It is necessary to try to understand this compactification also in concrete geometrical terms. What kind of a Riemann surface corresponds to the point at the infinity?

Here the situation is different from the general case. The reason really lies on the fact that the natural metric of a torus has curvature 0. Therefore each class of holomorphically homeomorphic torii has representatives whose area and/or diameter are either arbitrarily large or arbitrarily small. In the case of Riemann surfaces of genus > 1 the situation is completely different. By the Gauß–Bonnet theorem, the area of a Riemann surface of genus g, $g > 1$, is always $4\pi(g-1)$, i.e., the area depends only on the genus and not on the particular complex structure.

In order to understand properly the degeneration of torii it is, therefore, necessary to take particular representatives of each class of holomorphically isomorphic torii. One possibility is, for instance, to consider only those torii which have diameter < 2, i.e., which are of the form $\mathbf{C}/L_\tau$, $|\tau| < 1$.

This can be done, since for each τ' we can always find an element A of the elliptic modular group such that $|A(\tau')| < 1$. Considering these torii one can interpret the point at the infinity of M^1 *geometrically as a circle.*

The situation here is the following. As a sequence $\mathbf{C}/L_{\tau_n}$ approaches the infinity of the moduli space M^1, then we may suppose that the parameters τ_n are chosen in such a way that $\tau_n \to 0$, i.e., that the *radius of injectivity of* $\mathbf{C}/L_{\tau_n} \to 0$ as $n \to \infty$. This means that the limiting surface is a circle, and, as we let a point in M^1 approach the infinity, then the corresponding Riemann surfaces *collaps everywhere.* The limiting object is not anymore a surface but a manifold of real dimension 1.

The case of Riemann surfaces of genus > 1 is completely different as we will see in the proceeding sections. In the case of these hyperbolic Riemann surfaces collapsing can happen only in such a way that certain simple closed curves get replaced by points. The limiting structure is still a surface but it has finitely many simple singularities.

Let us next consider non–classical surfaces of genus 1. Such a (topological) surface Σ has, by the definition, the torus T^1 as its complex double, and there exists an orientation reversing involution $\sigma : T^1 \to T^1$ such that $\Sigma = T^1/\langle\sigma\rangle$. Such an involution induces a self–mapping σ^* of the Teichmüller space T^1 of the torus. Assume that σ_1 and σ_2 are different orientation reversing involutions of the torus. They induce self–mappings σ_1^* and σ_2^* of the Teichmüller space T^1, which usually are also different.

Since σ_1 and σ_2 are both orientation reversing, $\sigma_1 \circ \sigma_2$ is *orientation preserving.* Hence the induced mapping $(\sigma_1 \circ \sigma_2)^* = \sigma_2^* \circ \sigma_1^*$ belongs to the mapping class group Γ^1. Since $M^1 = T^1/\Gamma^1$ we conclude that both orientation reversing involutions σ_1 and σ_2 induce the same self–mapping of the moduli space M^1. This induced mapping is also an involution.

In order to find out what is the induced mapping at the level of the moduli space, consider the mapping

$$\sigma^* : U \to U,\ \sigma^*(\tau) = -\overline{\tau}.$$

A straightforward computation shows that this mapping $\sigma^* : T^1 \to T^1$ is induced by an orientation reversing involution $\sigma : T^1 \to T^1$ for which $T^1/\langle\sigma\rangle$ is an annulus. By Theorem 4.5.1 (on page 149), the Teichmüller space of an annulus can then be identified with the fixed–point set of the involution $\sigma^* : T^1 \to T^1$, i.e., with the imaginary axis.

Let

$$\tau : \mathbf{C} \to \mathbf{C},\ \tau(z) = \overline{z}.$$

On basis of the construction and the commutative diagram (5.3) we have: $j \circ \sigma^* = \tau \circ j$.

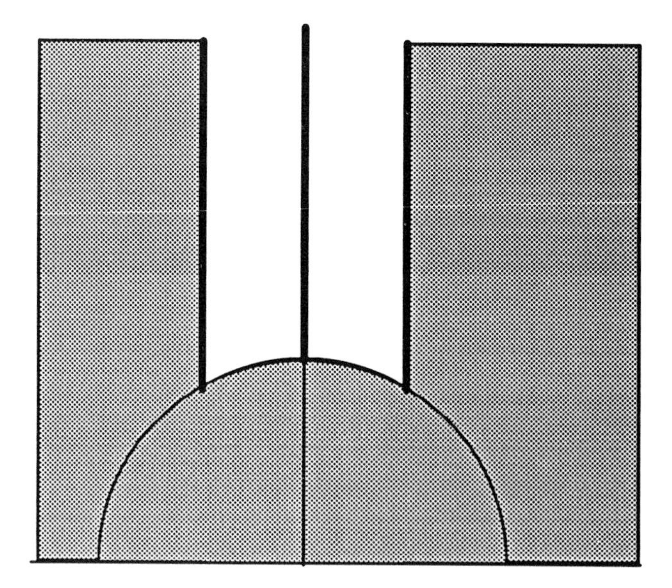

Figure 5.1: The unshaded part of the complex plane is a fundamental domain of the elliptic modular group. Those points of the fundamental domain for which $j : T^1 \to \mathbf{C}$ is real are indicated by a thick line.

Let $\pi : T^1 \to M^1$ denote the projection. We conclude now that be the above considerations,

$$\pi(T^1_{\sigma^*}) \subset \mathbf{R}.$$

Here $T^1_{\sigma^*}$ is the fixed–point set of the involution $\sigma^* : T^1 \to T^1$.

In order to see what kind of torii may have real moduli, let

$$(M^1)_{NC} = \{[X] \in M^1 \mid X \text{ has an antiholomorphic involution}\}.$$

The set $(M^1)_{NC}$ consists of ordinary isomorphism classes of Riemann surfaces which have a symmetry.

It is best to do the computations at the level of the the Teichmüller space. In figure 5.1 those points of a fundamental domain of the elliptic modular group, for which the function j takes real values, are indicated by a thick line. The elliptic modular function j takes real values also at many other points which do not belong to the fundamental domain of Figure 5.1. Detailed computations related to computing real values of the elliptic modular function can be found in the monograph of Norman Alling. [7, Chapter 9] We will skip the details here.

Direct computations show that $\{z \mid \operatorname{Im} z = \frac{1}{2}\}$ and $\{z \mid |z| = 1, \ \operatorname{Im} z > 0\}$ are both models for the Teichmüller space of the Möbius band while the imaginary axis is a model for Teichmüller spaces of the annulus and the Klein bottle. This implies that

$$(M^1)_{NC} = \mathbf{R}.$$

The above considerations are only technical and do not have any hidden difficulties. They can be interpreted in terms of algebraic geometry. Classical compact genus 1 Riemann surfaces are simply complex algebraic curves, i.e., they can be embedded into a projective space in which they are defined by a finite number of polynomial equations satisfying certain regularity conditions.

The moduli space M^1 is, therefore, the moduli space of genus 1 complex algebraic curves. Complex algebraic curves, which have an antiholomorphic involution, are, on the other hand, isomorphic to curves defined by *real polynomials*, i.e., they are *real algebraic curves.* The set $(M^1)_{NC}$ consists of complex isomorphism classes of real algebraic curves. By the above observations we have now:

Theorem 5.3.3 *The set of real points of the moduli space M^1, $M^1(\mathbf{R})$, consists of complex isomorphism classes of real algebraic curves.*

Theorem 5.3.3 is actually well known in algebraic geometry and easy to prove. For $j \neq 1728$, the j–invariant of the curve $y^2 = 4x^3 - ax - a$, $a = 27j/(j-1728)$, equals this given j. For $j = 1728$, take the curve $y^2 = 4x^3 - x$.

5.4 Stable Riemann surfaces

A main goal of this monograph is to understand, from a geometric point of view, the compactification given by Mumford and Bailey (cf. [65]) for the moduli space of smooth compact Riemann surfaces. To that end we have to extend our considerations to Riemann surfaces that are allowed to have singular points. In the proceeding definitions we follow the presentation of Bers ([12]).

Definition 5.4.1 *A* surface with nodes Σ *is a Hausdorff space whose every point has a neighborhood homeomorphic either to the open disk in the complex plane or to*

$$N = \{(z, w) \in \mathbf{C}^2 | zw = 0, |z| < 1, |w| < 1\}.$$

A point p of Σ is a node *if every open neighborhood of p contains a open set homeomorphic to N. Component of the complement of the nodes of Σ is a* part *of Σ. The* genus *of a compact surface with nodes Σ is the genus of the compact smooth surface obtained by thickening each node of Σ.*

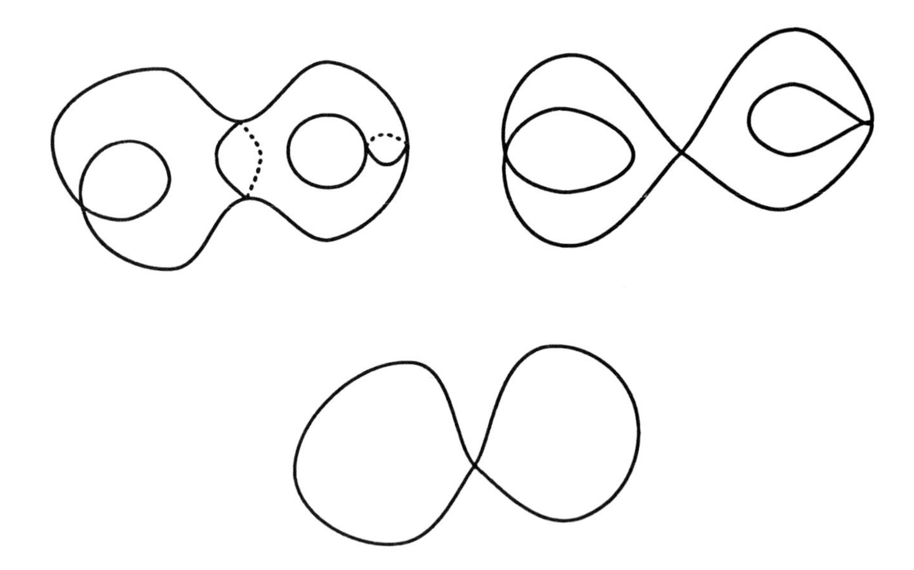

Figure 5.2: A stable surface, terminal stable surface and a non–stable surface with nodes.

A stable surface with nodes *is a compact surface with nodes whose every part has a negative Euler characteristic.* A stable Riemann surface with nodes *is a stable surface* Σ *together with a complex structure* X *for which each component of the complement of the nodes of* Σ *is obtained deleting a certain number* p_j *points from a compact Riemann surface of genus* g_j. *The stability condition means that*

$$2 - 2g_j - p_j < 0.$$

If X is a stable Riemann surface, then every part X_j of X is a hyperbolic Riemann surface, i.e., every X_j carries a canonical metric of constant curvature -1 which is obtained from the non–euclidean metric of the upper half–plane (or the unit disk) via uniformization. When we speak of lengths of curves on parts of a stable Riemann surface, we always refer to this canonical hyperbolic metric.

A stable surface Σ of genus g can have at most $3g - 3$ nodes. We say that Σ is *terminal* if it has this maximal number of nodes.

Definition 5.4.2 *A* strong deformation *with nodes* Σ_1 *onto a surface with nodes* Σ_2 *is a continuous surjection* $\Sigma_1 \to \Sigma_2$ *such that the following holds:*

- *the image of each node of* Σ_1 *is a node of* Σ_2,

- *the inverse image of a node of Σ_2 is either a node of Σ_1 or a simple closed curve on a part of Σ_1,*
- *the restriction of $\Sigma_1 \to \Sigma_2$ to the complement of the inverse image of the nodes of Σ_2 is an orientation preserving homeomorphism onto the complement of the nodes of Σ_2.*

Pairs of pants were defined (Definition 3.16.1 on page 117) as *closed* triply connected domains with boundary components. Allowing also pairs of pants that are degenerate in the sense that some of the boundary components are only points, one can decompose also stable Riemann surfaces into pairs of pants.

Let us next recall and extend the definitions of Section 3.16. Let Σ be a stable genus g surface. A *decomposition of Σ into pairs of pants* is an ordered collection

$$\mathcal{P} = (P_1, P_2, \ldots, P_{2g-2})$$

of disjoint pairs of pants on Σ such that:

- the union of the closures of the pairs of pants P_j covers the whole surface Σ.
- The intersection of the closures of any two pairs of pants P_i, P_j, $i \neq j$, is either empty or a union of nodes of Σ and of closed curves α on Σ.

It follows that all the nodes of Σ appear as boundary components of pairs of pants in any decomposition of Σ into pairs of pants.

If Σ is a terminal stable surface, then all the boundary components appearing in any decomposition of Σ into pairs of pants are nodes of Σ. If Σ is not terminal, then, in addition to the nodes, there will be a number of other boundary components which are simple closed curves on Σ. We call these nodes and curves *decomposing nodes and curves of $\mathcal{P}$*.

A decomposition $\mathcal{P} = (P_j)$ of Σ into pairs of pants is *oriented* if:

1. The set of boundary components of each pair of pants $P_j \in \mathcal{P}$ is ordered.
2. All decomposing curves are oriented.

If $\mathcal{P}$ is an oriented decomposition of Σ into pairs of pants, then we may speak of the first, second and third boundary component of any pair of pants belonging to $\mathcal{P}$. Furthermore, the ordering of the pairs of pants together with the ordering of the boundary components in the various pairs of pants induce an order in the set boundary components of the individual pairs of pants in the decomposition $\mathcal{P}$. Observe that each decomposing curve appears twice in this ordered set of boundary components of the pairs of pants.

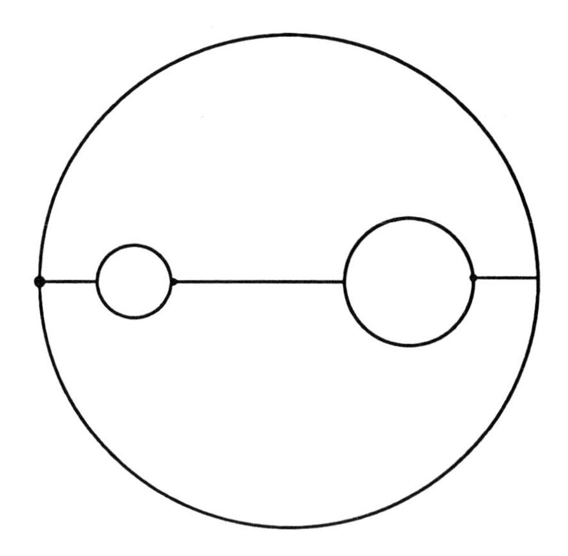

Figure 5.3: Base points on the boundaries of a pair of pants.

Let (P, d) be a pair of pants with ordered boundary components and with an intrinsic hyperbolic metric d. For a later construction it is necessary to associate *a base point* ζ_j to all boundary curves α_j of (P, d). That is done in the following way. Let $\gamma_{i,j}$ be the geodesic arc in (P, d) which joins the ith boundary component to the jth boundary component, i, $j = 1, 2, 3$, and is perpendicular to both of them. Such a geodesic arc is always uniquely defined. This is not hard to see and can be best shown considering the universal cover of the complex double of a pair of pants. The geodesic arcs $\gamma_{i,j}$ are then simply arcs on euclidean cirles perpendicular to the euclidean circles corresponding to the respective boundary components. For a notational convenience, define $\gamma_{3,4}$ setting $\gamma_{3,4} = \gamma_{3,1}$. The *base point* ζ_j of the boundary component α_j is the starting point of $\gamma_{j,j+1}$ on α_j.

Let X be a complex structure of Σ. Then $X = (\Sigma, X)$ is a stable Riemann surface. Each part of X carries a canonical hyperbolic metric.

An oriented decomposition of X into pairs of pants is called *geodesic* if every boundary curve of that decomposition is a geodesic curve on X. If $\mathcal{P}$ is any decomposition of X into pairs of pants, then we get always a geodesic oriented decomposition of X into pairs of pants by replacing each decomposing curve by the geodesic curve in its homotopy class.

Let $\mathcal{P}$ be any geodesic and oriented decomposition of X into pairs of pants. Let $\alpha_1, \ldots, \alpha_{3g-3}$ be the decomposing curves or points. Each curve (or point) α_j is either a boundary component of two different pairs of pants or appears twice as a boundary component of a single pair of pants.

Let ζ_j^k be the distinguished boundary point of the jth boundary component of the kth pair of pants of the decomposition $\mathcal{P}$, $j = 1,2,3$, $k = 1,\ldots,2g-2$. On each curve (or point) α_s, $s = 1,\ldots,3g-3$, there are exactly two points ζ_j^k. The ordering of the pairs of pants belonging to $\mathcal{P}$ and their respective boundary components gives us an ordering of these points ζ_j^k lying on one decomposing curve α_s. We conclude that on each α_s we have two distinguished points ξ_s^1 and ξ_s^2. These distinguished points are uniquely defined by the complex structure X.

Observe that strong deformations act on the set of oriented decompositions of a stable surface Σ into pairs of pants. More precisely, let $f : \Sigma' \to \Sigma$ be a strong deformation of stable surfaces and let $\mathcal{P}$ be an oriented decomposition of Σ into pairs of pants $P_1,\ldots,P_{2g-2}$. Then $f^*(\mathcal{P})$ is the decomposition of Σ' into pairs of pants $f^{-1}(P_j)$, $j = 1,2,\ldots,2g-2$. This is *the pull back* of the pants decomposition $\mathcal{P}$.

If $f : \Sigma' \to \Sigma$ is a strong deformation and $\mathcal{P}'$ is an oriented decomposition of Σ into pairs of pants such that each curve $f^{-1}\{\text{a node of } \Sigma\}$ is a decomposing curve (or point) of $\mathcal{P}$, then we can define the induced decomposition $f(\mathcal{P}')$ of Σ into pairs of pants. The pairs of pants of $f(\mathcal{P}')$ are images of pairs of pants in $\mathcal{P}'$ under the strong deformation f. If f is a homeomorphism, then the induced decomposition $f(\mathcal{P}')$ is defined for any decomposition $\mathcal{P}$.

Definition 5.4.3 *The set $\overline{M}^g$ of isomorphism classes of stable genus g Riemann surfaces is the* moduli space of stable genus g Riemann surfaces.

It turns out that $\overline{M}^g$ is a natural compactification of the moduli space M^g of smooth Riemann surface. This moduli space gets its topology from that of the Teichmüller space. The same construction cannot easily be carried out for $\overline{M}^g$. The reason is that the moduli space $\overline{M}^g$, which so far is only a set, does not have such a nice covering as the moduli space M^g. If one tries to extend geometrically the Teichmüller space so that it would cover all of the space $\overline{M}^g$, then one does not get a manifold anymore. One gets a rather complicated analytic space instead of a complex manifold. This approach, that was initiated by C. J. Earle and A. Marden ([27]) has recently been completed by F. Herrlich ([41]). In this monograph we will not follow that approach. We will resort to more concrete constructions that are based on the considerations of Chapter 3.

5.5 Fenchel–Nielsen coordinates

In this Section we recall the definition of the Fenchel–Nielsen coordinates for complex structures of stable topological surfaces.

Let Σ be a stable topological surface of genus g. Fix an oriented decomposition $\mathcal{P}$ of Σ into pairs of pants. Let X be a complex structure of Σ. Each homotopy class of closed curves on a hyperbolic Riemann surface always contains a unique geodesic curve. When defining parameters for a complex structure X we first replace the decomposing curves of $\mathcal{P}$ with the unique geodesic curves homotopic to the original decomposing curves. In view of our applications we may, without restricting the generality, suppose that the decomposing curves of $\mathcal{P}$ are already geodesic curves, i.e., we may limit our consideration to complex structures $X \in \mathcal{M}(\mathcal{P})$.

Let $\alpha_1, \ldots, \alpha_{3g-3}$ be the oriented decomposing curves of the pants decomposition $\mathcal{P}$. Recall that any $X \in \mathcal{M}(\mathcal{P})$ defines two distinguished points ξ_j^1 and ξ_j^2 on each α_j. Let s_j denote the distance from ξ_j^1 to ξ_j^2 measured to the positive direction of α_j.

In $\mathcal{M}(\mathcal{P})$ we can define the functions ℓ_j and θ_j, $j = 1, 2, \ldots, 3g-3$, setting

$$\begin{aligned} \ell_j &= \text{the length of } \alpha_j \\ \theta_j &= 2\pi s_j/\ell_j \text{ if } \ell_j > 0 \\ \theta_j &= 0 \text{ if } \ell_j = 0 \end{aligned} \tag{5.4}$$

It is clear that X and $X' \in \mathcal{M}(\mathcal{P})$ are isomorphic complex structures if $\ell_j(X) = \ell_j(X')$ and $\theta_j(X) = \theta_j(X')$ for all $j = 1, 2, \ldots, 3g-3$. A necessary and sufficient condition for X and X' to be isomorphic is that there exists a homeomorphism $f : X \to X'$ and a decomposition $\mathcal{P}'$ of X' into pairs of pants such that the following holds:

1. $\mathcal{P} = f^*(\mathcal{P}')$.

2. Let ℓ_j' and θ_j' be the coordinates of X' with respect to $\mathcal{P}'$ which correspond to the coordinates ℓ_j and θ_j of X with respect to $\mathcal{P}$. Then $\ell_j = \ell_j'$ and $\theta_j = \theta_j'$ for all $j = 1, 2, \ldots, 3g-3$.

Definition 5.5.1 *The coordinates ℓ_j and θ_j are the* Fenchel–Nielsen coordinates, *the coordinates ℓ_j are called the* length coordinates *or the* Fenchel–Nielsen length coordinates *and the coordinates θ_j are called the* gluing angles *or the* Fenchel–Nielsen gluing angles.

The above definition for Fenchel–Nielsen coordinates is classical. In some considerations it would be a technical simplification to replace the Fenchel–Nielsen gluing angle θ_j by the product $\theta_j' = \ell_j\theta_j$. We call the coordinates

ℓ_j and θ'_j the *modified Fenchel–Nielsen coordinates.* For a clearly written account of the Fenchel–Nielsen coordinates see [30].

Let us take a closer look at the Fenchel–Nielsen coordinates of a *smooth* Riemann surface. So we suppose now that Σ is an ordinary topological genus g surface (without nodes). Let $f_1, f_2, \ldots, f_{3g-3}$ denote the left Dehn twists around the decomposing curves of the pants decomposition $\mathcal{P}$. Let $\Gamma(\mathcal{P})$ be the subgroup of the modular group $\Gamma(\Sigma)$ freely generated by the elements f_j^*, $j = 1, 2, \ldots, 3g-3$. Then every element of $\Gamma(\Sigma)$ is of infinite degree and $\Gamma(\Sigma)$ does not contain any finite subgroups save the trivial subgroup consisting only of the identity.

The following result is obvious by the definitions:

Lemma 5.5.1 *The Fenchel–Nielsen coordinates establish a bijection*

$$T(\Sigma)/\Gamma(\mathcal{P}) \to (\mathbf{R}_+ \times [0, 2\pi))^{3g-3}, \; [X] \mapsto (\ell_1, \theta_1, \ldots, \ell_{3g-3}, \theta_{3g-3}). \quad (5.5)$$

Observe that in the above lemma we may visualize the product $\mathbf{R}_+ \times [0, 2\pi)$ as the complex plane punctured at the origin. Consequently, the mapping (5.5) can be interpreted also as an injective mapping

$$F_{\mathcal{P}} : T(\Sigma)/\Gamma(\mathcal{P}) \to \mathbf{C}^{3g-3}.$$

The image of $T(\Sigma)$ under this mapping consists of all points of $\mathbf{C}^{3g-3}$ with all coordinates non–zero.

Let $\pi : T(\Sigma) \to T(\Sigma)/\Gamma(\mathcal{P})$ denote the projection.

Lemma 5.5.2 *The mapping*

$$F_{\mathcal{P}} \circ \pi : T(\Sigma) \to \mathbf{C}^{3g-3}$$

is real analytic.

Proof. By Theorem 4.9.3 (on page 160), the geodesic length functions are real analytic. The Fenchel–Nielsen length coordinates are therefore real analytic.

To show the real analyticity of the Fenchel–Nielsen gluing angles, recall the characterization of the complex structure of the Teichmüller space $T(\Sigma)$ as given on page 148. In view of that characterization we have to show that the Fenchel–Nielsen gluing angles vary real analytically as we deform the complex structure by a Beltrami differential μ. That is straightforward and can be seen in the following fashion.

Let α be one of the decomposing geodesics of the pants decomposition $\mathcal{P}$. Denote the corresponding gluing angle by θ. We show that θ varies real analytically as we deform the Riemann surface X by a Beltrami differential μ.

To that end, let let β and γ be the two other decomposing geodesics of $\mathcal{P}$ such that the end–points ξ_1 and ξ_2 of the geodesic arcs perpendicular to α and β or to α and γ, respectively, are the two distinguished points of the decomposing geodesic curves α.

Use the uniformization to express X as U/G for a Fuchsian group G acting in the upper half–plane. Then we may choose hyperbolic Möbius transformations g, h and h' such that

- the axis of g covers α,
- the axis of h covers β,
- the axis of h' covers γ.

We may, furthermore, suppose that the configuration is that of Figure 5.4 were also the geodesic arcs defining the distinguished points is shown. In that figure we have x_1 and x_2 are points lying over the distinguished points ξ_1 and ξ_2 of α. Depending on the choice of the orientation of α, the Fenchel–Nielsen gluing angle θ associated to α is now simply either $\theta = 2\pi d_U(x_1, x_2)/\ell_\alpha$ or $\theta = 2\pi(1 - d_U(x_1, x_2)/\ell_\alpha)$.

It suffices to show, therefore, that the hyperbolic distance $d_U(x_1, x_2)$ of x_1 and x_2 in U varies real analytically as we deform the group G by a Beltrami differential μ of the group.

This deformation of G by μ means the following process. Assume first that μ is a Beltrmai differential of the group G. Recall the definition of Beltrami differentials of groups, Definition 4.2.3 on page 143. Let $f^\mu : \overline{U} \to \overline{U}$ be the uniquely defined quasiconformal mapping which has the complex dilatation μ in U and fixes the points 0, 1 and ∞. The *μ–deformation* of G is then simply the Fuchsian group $G_\mu = f^\mu G (f^\mu)^{-1}$.

An important theorem of Lars V. Ahlfors([2]) and Lipman Bers implies that the values of the mapping f^μ depend *holomorphically* on the complex dilatation μ (cf. Theorem 2.4.6 on page 65, or [60, Theorem 3.1, p. 69 and §V.5.1]). This implies also that $d_U(x_1, x_2)$ depends real analytically on μ.

5.6 Topology for the moduli space of stable Riemann surfaces

Our aim is to define a topology for the space of isomorphism classes of stable compact Riemann surfaces of a given genus g, $g > 1$, in such a way that it is possible to understand the degeneration of smooth Riemann surfaces in a concrete geometric fashion. That can be achieved by the Fenchel–Nielsen coordinates. In this Section we first give a definition of the topology and

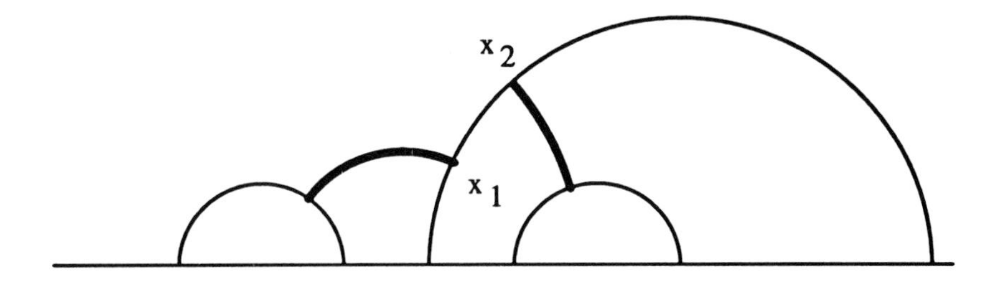

Figure 5.4: Geodesic arcs defining the distinguished points x_1 and x_2.

then show how to interpret the degeneration of smooth Riemann surfaces in this topology.

Let ϵ and δ be positive numbers. We say that the point $[Y] \in \overline{M}^g$ belongs to the (ϵ, δ)–neighborhood $U_{\epsilon,\delta}([X])$ if and only if there exists a decomposition $\mathcal{P}$ of X into pairs of pants such that the following holds:

> There exists a strong deformation $f : Y \to X$ such that the Fenchel–Nielsen coordinates $\ell'_1, \theta'_1, \ldots, \ell'_{3g-3}, \theta'_{3g-3}$ of Y with respect to $f^*(\mathcal{P})$ satisfy:
>
> 1. $|\ell_j - \ell'_j| < \epsilon$ for all $j = 1, 2, \ldots, 3g-3$.
> 2. For all values of j, if $\ell_j > 0$ then $|\theta_j - \theta'_j| < \delta$.

Here we have used the notation $f^*(\mathcal{P})$ to indicate the pants decomposition of Y whose pairs of pants are inverse images of pairs of pants of the decomposition $\mathcal{P}$ of X, and ℓ_j and θ_j are the Fenchel–Nielsen coordinates of X with respect to $\mathcal{P}$.

The sets $U_{\epsilon,\delta}([X])$ form a basis for the topology of $\overline{M}^g$.

Theorem 5.6.1 *The set of isomorphism classes of smooth genus g Riemann surfaces, M^g, is dense in the moduli space of stable compact genus g Riemann surfaces, $\overline{M}^g$, which is a connected Hausdorff space.*

Proof. By the construction of the topology, M^g is dense in $\overline{M}^g$. This is clear, since each $U_{\epsilon,\delta}([X])$, $\epsilon > 0$, neighborhood of a stable Riemann surface contains smooth Riemann surfaces.

Proof for the fact that this topology is a Hausdorff topology is rather obvious and is left to the reader. The connectedness of $\overline{M}^g$ follows from that of M^g (Lemma 5.2.1).

The following result follows directly from the definition and from Lemma 5.5.2.

Lemma 5.6.2 *Consider the Teichmüller space T^g of compact genus g Riemann surfaces, $g > 1$, equipped with the usual topology. The projection $T^g \to \overline{M}^g$ is a continuous mapping.*

5.7 Compactness theorem

In this section we show that the moduli space $\overline{M}^g$, $g > 1$, is compact. In view of Theorem 5.6.1 it is enough to show the following:

Theorem 5.7.1 *The closure of M^g in $\overline{M}^g$ is compact.*

Proof. Let $([(\Sigma, X)])$ be an infinite sequence of points of $M^g \subset \overline{M}^g$. It suffices to show that there exists a subsequence $([(\Sigma, X_{n_k})])$ that converges in $\overline{M}^g$.

By Theorem 3.19.1 on page 131 we can always find a decomposition $\mathcal{P}_n$ of (Σ, X_n) into pairs of pants by simple closed geodesic curves of length $< 21g$. This is the key point in our argument.

There are only finitely many topologically different decompositions of the surface Σ into pairs of pants. Therefore we may — by passing to a subsequence — assume that there is a fixed decomposition $\mathcal{P}$ of Σ into pairs of pants and orientation preserving homeomorphisms $f_n : \Sigma \to \Sigma$ such that $f_n(\mathcal{P}_n) = \mathcal{P}$ for each n.

Let Y_n be that complex structure of Σ for which the mapping

$$f_n : (\Sigma, X_n) \to (\Sigma, Y_n)$$

is holomorphic.

Let ℓ_j^n, θ_j^n, $j = 1, 2, \ldots, 3g - 3$, be the Fenchel–Nielsen coordinates of Y_n with respect to the pants decomposition $\mathcal{P}$. Let $\alpha_1, \ldots, \alpha_{3g-3}$ be the decomposing curves of the pants decomposition $\mathcal{P}$. Then $\ell_j^n = \ell_{\alpha_j}(Y_n)$ is the length of the simple closed geodesic curve freely homorotopic to the closed curve α_j on the Riemann surface Y_n.

By the choice of the pants decompositions $\mathcal{P}_n$ of the Riemann surfaces (Σ, X_n) and by the definition of the complex structure Y_n we have now, for $M = 21g$:

$$\ell_j^n = \ell_{\alpha_j}(Y_n) < M$$

for each j and n. We have also $0 \leq \theta_j^n < 2\pi$. Therefore — by passing to a subsequence again — we may assume that all the sequences $\ell_j^1, \ell_j^2, \ell_j^3, \ldots$ and $\theta_j^1, \theta_j^2, \theta_j^3, \ldots$ converge. Let $\ell_j = \lim_{n\to\infty} \ell_j^n$ and $\theta_j = \lim_{n\to\infty} \theta_j^n$, $j = 1, 2, \ldots, 3g - 3$.

We deform next the surface Σ in the following fashion. If $\ell_j = 0$ then we replace the decomposing curve α_j of the pants decomposition $\mathcal{P}$ by a node. Do that for each j with $\ell_j = 0$. That construction yields a stable surface Σ^* of genus g. The identity mapping $\Sigma \to \Sigma$ induces a strong deformation

$$f : \Sigma \to \Sigma^*. \tag{5.6}$$

On Σ^* define a complex structure Y^* using the Fenchel–Nielsen coordinates ℓ_j and θ_j with respect to the decomposition $\mathcal{P}$ of Σ^* into pairs of pants. That is done in the following way. Using Lemma 4.1.1 we define first, in each pair of pants P_j, the complex structure Y^* by the limits ℓ_j, $j = 1, 2, \ldots, 3g-3$, of the Fenchel–Nielsen length coordinates ℓ_j^n associated to Y_n.

Next we glue these complex structures of the various pairs of pants together to form a global complex structure of the surface Σ^*. That we do by the identification pattern given by the original decomposition $\mathcal{P}$ of Σ^* and the gluing angles given by the limits $\theta_j = \lim_{n\to\infty} \theta_j^n$. That gives us the complex structure[1] Y^* of Σ^*. From the construction it follows then immediately that

$$[(\Sigma, X_n)] = [(\Sigma, Y_n)] \to [(\Sigma^*, Y^*)] \quad \text{as} \quad n \to \infty$$

in $\overline{M^g}$. This proves the theorem.

5.8 Symmetric Riemann surfaces and real algebraic curves

Next we turn our attention to non–classical compact Riemann surfaces. Their study has its motivation in the fact that non–classical Riemann surfaces are simply complex algebraic curves defined by real polynomials (cf. [8]). This observation justifies the following definition.

Definition 5.8.1 *The space of isomorphism classes of smooth non–classical Riemann surfaces $M^g_{\mathbf{R}}$ of genus g is called the* moduli space of real algebraic curves of genus g. *Alternatively we call the space $M^g_{\mathbf{R}}$* moduli space of smooth non–classical genus g Riemann surfaces.

[1] This complex structure is not yet uniquely defined by the above data. It is defined up to full Dehn twists around the decomposing curves of the pants decomposition $\mathcal{P}$. In spite of this ambiguity, the point $[Y^*] \in \overline{M^g}$ *is uniquely defined,* and this ambiguity does not cause any problems.

The moduli space of smooth genus g real curves is, for the time being, defined only as a set. Below we will define a topology for this set. Before doing that recall that by Corollary 3.4.2 (on page 78) there are

$$\lfloor\frac{3g+4}{2}\rfloor$$

different topological types of non–classical genus g Riemann surfaces. This implies immediately that, in any reasonable topology, the moduli space $M_{\mathbf{R}}^g$ will not be connected, since you cannot change the topological type of a non–classical Riemann surface without doing some violence to it.

Recall that by passing to the complex double X^c of a non–classical Riemann surface X we may view X as a classical Riemann surface with a symmetry, i.e., with an antiholomorphic involution $\sigma : X^c \to X^c$. (For the definition of the complex double see page 77.)

Definition 5.8.2 *Let (X, σ) and (Y, τ) be stable Riemann surfaces with symmetries σ and τ. We say that (X, σ) and (Y, τ) are* isomorphic *(or* isomorphic over $\mathbf{R}$*) if there exists a holomorphic homeomorphism $f : X \to Y$ such that $f \circ \sigma = \tau \circ f$.*

It is now clear that the moduli space $M_{\mathbf{R}}^g$ can be viewed as the set of isomorphism classes of symmetric and smooth genus g Riemann surfaces. In order to understand the natural compactification of $M_{\mathbf{R}}^g$ it is best to view it as the space of isomorphism classes of symmetric surfaces. The reason is the following observation:

> Let $[X_n] \in M_{\mathbf{R}}^g$ be a infinite sequence of isomorphism classes of non–classical smooth genus g Riemann surfaces which does not converge in $M_{\mathbf{R}}^g$. It may happen that all non–classical Riemann surfaces X_n have two boundary components such that the length of one of them (measured in the intrinsic metric) converges to 0 as $n \to \infty$ and the length of the other one converges to ∞ as $n \to \infty$.

This implies that, as $n \to \infty$, the non–classical Riemann surfaces X_n degenerate in two completely different manners. It is also rather hard to understand what this means geometrically. We will, therefore, from now on interpret $M_{\mathbf{R}}^g$ always as the set of isomorphism classes of symmetric smooth genus g Riemann surfaces.

Definition 5.8.3 *The set $\overline{M_{\mathbf{R}}^g}$ of isomorphisms classes of symmetric stable genus g Riemann surfaces is called the* moduli space of symmetric stable genus g Riemann surfaces. *Alternatively it is called the* moduli space of stable real algebraic curves of genus g.

Just like in the classical case, the moduli space $\overline{M}^g_{\mathbf{R}}$ is a compactification of the moduli space $M^g_{\mathbf{R}}$. In order to understand this, we have, of course, first to define a topology for these both moduli spaces. The moduli space $M^g_{\mathbf{R}}$ is naturally a subset of $\overline{M}^g_{\mathbf{R}}$. We define a topology for $\overline{M}^g_{\mathbf{R}}$ which then induces a topology for $M^g_{\mathbf{R}}$.

Here we repeat the construction given for the moduli space $\overline{M}^g$.

Let ϵ and δ be positive numbers. We say that the point $[(Y,\tau)] \in \overline{M}^g_{\mathbf{R}}$ belongs to the (ϵ,δ)–neighborhood $U_{\epsilon,\delta}([(X,\sigma)])$ of a point $[(X,\sigma)]$ if and only if there exists a decomposition $\mathcal{P}$ of X into pairs of pants such that the following holds:

1. $\sigma(\mathcal{P}) = \mathcal{P}$.

2. There exists a strong deformation $f : Y \to X$ such that the following holds:

   (a) $f \circ \tau = \sigma \circ f$.

   (b) Fenchel–Nielsen coordinates $\ell'_1, \theta'_1, \ldots, \ell'_{3g-3}, \theta'_{3g-3}$ of Y with respect to $f^*(\mathcal{P})$ satisfy:

      i. $|\ell_j - \ell'_j| < \epsilon$ for all $j = 1, 2, \ldots, 3g-3$.

      ii. For all values of j, if $\ell_j > 0$ then $|\theta_j - \theta'_j| < \delta$.

Here we have used the notation $f^*(\mathcal{P})$ to indicate the pants decomposition of Y whose pairs of pants are inverse images of pairs of pants of the decomposition $\mathcal{P}$ of X, and ℓ_j and θ_j are the Fenchel–Nielsen coordinates of X with respect to $\mathcal{P}$.

The sets $U_{\epsilon,\delta}([X])$ form a basis for the topology of $\overline{M}^g$.

5.9 Connectedness of the real moduli space

In this Section we show that the above defined moduli space for stable symmetric Riemann surfaces, i.e., the real moduli space of stable real algebraic curves, is connected. The proof presented here is a modification of the arguments of [76], see also [78]. The connectedness of the moduli space $\overline{M}^g_{\mathbf{R}}$ is actually an old problem. It was considered already by Felix Klein in [48, Page 8]. We start with an almost obvious result.

Theorem 5.9.1 *The moduli space $\overline{M}^g_{\mathbf{R}}$, $g > 1$, is a Hausdorff space and $M^g_{\mathbf{R}}$ is dense in $\overline{M}^g_{\mathbf{R}}$.*

Proof. It is immediate that $\overline{M}^g_{\mathbf{R}}$ is a topological Hausdorff space. To show that $M^g_{\mathbf{R}}$ is dense in $\overline{M}^g_{\mathbf{R}}$ we have to show that we can thicken all the nodes of a symmetric stable Riemann surface (X, σ) in a fashion which is compatible with the involution σ. That is, indeed, possible. Details are similar to those of (a part of) the proof of Theorem 5.9.3.

In order to study $M^g_{\mathbf{R}}$ more closely, assume that integers g, n and k satisfy:

- $g > 1$.
- If $k = 0$ then $n \equiv g + 1 \pmod 2$ and $n > 0$.
- If $k = 1$ then $0 \leq n \leq g$.

Then, by Theorem 3.4.1 (on page 78), there are symmetric Riemann surfaces (X, σ) of genus g such that the number of the components of the fixed–point set of σ equals n and $k = 2 - \#(\text{components of } X \setminus X_\sigma)$.

Let

$$V(g, n, k) = \{[(X, \sigma)] \in M^g_{\mathbf{R}} \mid \text{ invariants of } (X, \sigma) \text{ equal } g, n \text{ and } k\}.$$

It is obvious, by the definitions, that the sets $V(g, n, k)$ are disjoint subsets of $\overline{M}^g_{\mathbf{R}}$ for different values of the invariants n and k.

Definition 5.9.1 *The set* $V(g, n, k)$ *is the* moduli space *of smooth symmetric Riemann surfaces of topological type* (g, n, k).

Lemma 5.9.2 *The space* $V(g, n, k)$ *is connected,* $g > 1$.

Proof. Let Σ be a topological compact non–classical surface of type (g, n, k)[2]. By Theorem 4.4.1 (on page 147) the Teichmüller space $T(\Sigma)$ of the surface Σ is connected. This implies the connectedness of $V(g, n, k)$ in the following fashion.

Let Σ^c be the complex double of Σ. Recall the constructions related to Theorem 4.5.1. The complex double Σ^c carries an orientation reversing involution $\sigma : \Sigma^c \to \Sigma^c$ such that $\Sigma = \Sigma^c/\langle\sigma\rangle$. This involution induces an involution $\sigma^* : T(\Sigma^c) \to T(\Sigma^c)$. By means of the projection $\pi : \Sigma^c \to \sigma$ we constructed a continuous and injective mapping $\pi^* : T(\Sigma) \to T(\Sigma^c)$. By Theorem 4.5.1,

$$\pi^*(T(\Sigma) = T(\Sigma^c)_{\sigma^*}. \tag{5.7}$$

Equation (5.7) implies now that $T(\Sigma^c)_{\sigma^*}$ is connected.

Let $N(\sigma^*)$ be the *normalizer* of σ^* in $\Gamma(\Sigma^c)$. It consists of all elements of $\Gamma(\Sigma^*)$ that commute with σ^*. Then elements of $N(\sigma^*)$ map $\pi^*(T(\Sigma)) =$

[2]The surface Σ has n boundary components, is orientable if $k = 0$ and non–orientable otherwise and the ordinary genus of the complex double of Σ is g.

$T(\Sigma^c)_{\sigma^*}$ onto itself. Since $\pi^*(T(\Sigma))$ is connected, it follows that $V(g,n,k) = \pi^*(T(\Sigma))/N(\sigma^*)$ is also connected.

Theorem 5.9.3 *The moduli space $\overline{M_{\mathbf{R}}^g}$ of symmetric stable genus g Riemann surfaces, $g > 1$, is a connected Hausdorff space.*

Proof. By Theorem 5.9.1, $M_{\mathbf{R}}^g$ is dense in $\overline{M_{\mathbf{R}}^g}$. Hence it suffices to show that the closure of $M_{\mathbf{R}}^g$ in $\overline{M_{\mathbf{R}}^g}$ is connected.

We will achieve this in two steps:

- We show first that the union of the closures of components $V(g,n,1)$ is connected by showing that the closure of each $V(g,n,1)$, $n > 0$, intersects the closure of $V(g,0,1)$.
- Next we show that the closures of $V(g,n,0)$ and $V(g,n-1,1)$ always intersect for all possible values of n.

Consider now a component $V(g,n,1)$, $n > 0$. To show that the closures of $V(g,n,1)$ and $V(g,0,1)$ intersect we construct a point $[(X,\sigma)]$ that lies in the closures of both of them.

To that end, let X_0 be the Riemann sphere punctured at the $g+1$ points $0, 1, 2, \ldots, g$. Let X_0^- be the complex conjugate of X_0, i.e., if z is a local coordinate of X_0, then $\bar{z}$ is that of X_0^-.

Let X be the stable Riemann surface obtained by identifying the punctures $0, 1, \ldots, g$ of X_0 with those of X_0^-. The identity mapping induces an antiholomorphic mapping $X_0 \to X_0^-$ which, in turn, induces a symmetry $\sigma : X \to X$. We conclude that (X,σ) is a stable genus g Riemann surface.

The stable Riemann surface X has two parts, they correspond to the punctured spheres X_0 and X_0^-. For notational convenience, call these parts X_1 and X_2. The symmetry σ maps X_1 onto X_2.

Next we show that the point $[(X,\sigma)]$ lies in the closures of $V(g,0,1)$ and $V(g,n,1)$. To that, let $\epsilon > 0$ and $\delta > 0$ be arbitrary. We constructs a point of $V(g,0,1)$ and another one of $V(g,n,1)$ which both lie in the $U_{\epsilon,\delta}([(X,\sigma)])$ neighborhood of the point $[(X,\sigma)]$.

To that end, take first $g-1$ pairs of pants $P_1, P_2, \ldots, P_{g-1}$ such that all boundary geodesics have length $\epsilon/2$. identify a boundary component of P_j with a boundary component of P_{j+1} for all indices $j = 1, \ldots, g-2$. Do the identifications in such a manner that the corresponding base points always agree. In this way one obtains a Riemann surface Y' with $(g-1)+2 = g+1$ boundary components.

Let Y'' be the complex conjugate of Y'. As above, we form the Riemann surface Y by identifying the boundary points of Y' with the corresponding points of Y''. The identity mapping of Y' induces then an antiholomorphic mapping $Y' \to Y''$ which, in turn, induces a symmetry $\sigma : Y \to Y$. The Riemann surface Y is, of course, smooth. The point $[(Y\sigma)]$ lies in the component $V(g, g+1, 0)$.

Next we deform the complex structure of Y in such a way that we get the desired points of $V(g, 0, 1)$ and of $V(g, n, 1)$. This is a delicate part of the argument and is based on the considerations of Section 3.7 (starting on page 93).

Let $\alpha_1, \alpha_2, \ldots, \alpha_{g+1}$ be the closed geodesic curves of Y corresponding to the boundary geodesics of the Riemann surface Y'. Assume that the curves α_j are oriented in such a way that they are positively oriented[3] as boundary curves of Y'. Let $k = g+1-n$, and let Y_k be the Riemann surface obtained from the Riemann surface Y be performing left Dehn twists along the curves $\alpha_1, \ldots, \alpha_k$. For the definition of the left Dehn twist see page 95.

Recall now the construction of the mapping f_α given in Definition 3.7.1 on page 94. Considering carefully the construction one concludes that

> the homotopy class of the mapping $f_{\alpha_1} \circ f_{\alpha_2} \circ \cdots \circ f_{\alpha_k} \circ \sigma$ contains an antiholomorphic involution $\tau : Y_k \to Y_k$.

It is immediate[4], by the definition of f_{α_j}, that the symmetric Riemann surface (Y, τ) is of type $(g, n, 1)$, i.e., that $[(Y, \tau)] \in V(g, n, 1)$. It follows, also directly from the definitions, that $[(Y, \tau)] \in U_{\epsilon,\delta}([(X, \sigma)])$. This argument shows that the point $[(X, \sigma)]$ lies in the closure of $V(g, n, 1)$.

Replacing, in the above construction, k by $g+1$ one shows that $[(X, \sigma)]$ lies also in the closure of $V(g, 0, 1)$. This completes the proof of the first part.

Observe that in the above we actually showed that the point $[(X, \sigma)]$ lies in the closures of the components $V(g, n, 1)$, $V(g, 0, 1)$ *and* $V(g, g+1, 0)$, i.e., that also the closures of $V(g, g+1, 0)$ and $V(g, 0, 1)$ intersect.

In order to show that the closure of $V(g, 0, 1)$ intersects the closures of all the components $V(g, n, 0)$, $0 < n \leq g+1$, $n \equiv g+1 \pmod 2$, we repeat the above argument by replacing the punctured sphere Y' by a Riemann surface of genus $(g+1-n)/2$ from which n disks (with disjoint closures) have been removed. (Here we repeat the construction of Section 3.7 starting on page 93.) Details are, word by word, same as in the above considerations. This completes the proof.

The main point in the above proof is that the involutions $\tau \approx f_{\alpha_1} \circ f_{\alpha_2} \circ \cdots \circ f_{\alpha_k} \circ \sigma$ and σ are not homotopic to each other. The composition $\tau \circ \sigma^{-1}$

[3] Any orientation will do. For technical convenience we have to fix *some* orientation.

[4] Details here are exactly the same as in the considerations following Definition 3.7.1 on page 94.

is, of course, homotopic to the product of Dehn twists $f_{\alpha_1} \circ f_{\alpha_2} \circ \cdots \circ f_{\alpha_k}$. As a self mapping of the smooth surface Σ this product of Dehn twists is not homotopic to the identity and is of infinite degree.

If one deforms the surface Σ in such a way that the closed curves $\alpha_1, \ldots, \alpha_k$ get pinched to nodes, then the mapping $f_{\alpha_1} \circ f_{\alpha_2} \circ \cdots \circ f_{\alpha_k}$ induces a self-mapping of this deformed surface which *is* homotopic to the identity. It only turns the nodes around.

5.10 Compactness of the real moduli space

We use Theorem 5.9.1 and study the moduli space $\overline{M}^g_{\mathbf{R}}$ as the union of the closures of the parts:

$$V(g,n,k) = \{[(X,\sigma)] \in \overline{M}^g_{\mathbf{R}} | X \text{ smooth}, \ n(\sigma) = n, \ k(\sigma) = k\}.$$

We will first show that the closure of each $V(g,n,k)$ is compact in $\overline{M}^g_{\mathbf{R}}$.

The proof of the compactness is an extension of the arguments presented in Section 5.7 for the classical case and relies on Theorem 3.19.1 (on page 131). Observe especially, that the following result follows immediately from the definitions.

Lemma 5.10.1 *The projection*

$$\overline{M}^g_{\mathbf{R}} \to \overline{M}^g, \ [(X,\sigma)] \mapsto [X] \tag{5.8}$$

is continuous.

We will have to deal with several different symmetries of the surface Σ. To make this distinction clear we write sometimes (Σ, X, σ) to denote the symmetric Riemann surface $X = (\Sigma, X)$ together with the symmetry $\sigma : (\Sigma, X) \to (\Sigma, X)$ which is then an antiholomorphic involution.

Theorem 5.10.2 *The closure of* $V(g,n,k)$ *in* $\overline{M}^g_{\mathbf{R}}$ *is compact.*

Proof. Let $\sigma : \Sigma \to \Sigma$ be an orientation reversing involution, $k(\sigma) = k$ and $n(\sigma) = n$. Let $([(\Sigma, X_n, \sigma)])$ be an infinite sequence of points of $V(g,n,k)$ in $\overline{M}^g_{\mathbf{R}}$. It suffices to show that there exists a subsequence

$$([(\Sigma, X_{n_k}, \sigma)])$$

that converges in $\overline{M}^g_{\mathbf{R}}$.

We shall, at various stages of the proof pass from a sequence to its subsequence. To keep the notation as simple as possible we use the same notation for a sequence and its suitable subsequence when there is no danger of confusion.

Use first Theorem 3.19.1 (page 131) to find, for each index n, a decomposition $\mathcal{P}_n$ of Σ into pairs of pants in such a way that each decomposing curve α_j^n of each pants decomposition $\mathcal{P}_n$ has length $< 21g$ on (Σ, X_n). By Theorem 3.19.1 we can furthermore choose these pants decompositions $\mathcal{P}_n$ in such a way that, for each n, $\sigma(\mathcal{P}_n) = \mathcal{P}_n$.

There are only finitely many topologically different decompositions of the surface Σ into pairs of pants. Therefore we may — by passing to a subsequence — assume that there is a fixed decomposition $\mathcal{P}$ of Σ into pairs of pants and orientation preserving homeomorphisms $f_n : \Sigma \to \Sigma$ such that $f_n(\mathcal{P}_n) = \mathcal{P}$ for each n.

Let Y_n be that complex structure of Σ for which the mapping $f_n : (\Sigma, X_n) \to (\Sigma, Y_n)$ is holomorphic. Let $\tau_n = f_n \circ \sigma \circ f_n^{-1}$. Each mapping $\tau_n : (\Sigma, Y_n) \to (\Sigma, Y_n)$ is then an antiholomorphic involution. Furthermore, $\tau_n(\mathcal{P}) = \mathcal{P}$ for each n.

Up to Dehn twists around the decomposing curves of the pants decomposition $\mathcal{P}$ there are only finitely many different involutions τ_n that map $\mathcal{P}$ onto itself. Therefore we may assume — by passing again to a subsequence and choosing the mappings f_n in a suitable way — that all the involutions τ_n agree. Let $\tau = \tau_n$ for all values of n.

After these choices we have a decomposition $\mathcal{P}$ of Σ into pairs of pants and representatives (Σ, Y_n, τ) for the points $[(\Sigma, X_n, \sigma)]$ in $\overline{M_{\mathbf{R}}^g}$ such that the following holds:

1. $\tau(\mathcal{P}) = \mathcal{P}$.
2. Each $\tau : (\Sigma, Y_n) \to (\Sigma, Y_n)$ is an antiholomorphic involution.
3. Each decomposing curve α_j, $j = 1, 2, \ldots, 3g-3$, of the pants decomposition $\mathcal{P}$ is of length $< 21g$ on the hyperbolic Riemann surface (Σ, Y_n).

Let ℓ_j^n, θ_j^n, $j = 1, 2, \ldots, 3g-3$, be the Fenchel–Nielsen coordinates of Y_n with respect to the pants decomposition $\mathcal{P}$. Then by the above mentioned property 3, we have $\ell_j^n < 21g$ for each j and n. Also we have $0 \leq \theta_j^n < 2\pi$. Therefore — by passing again to a subsequence — we may assume that all the sequences $\ell_j^1, \ell_j^2, \ell_j^3, \ldots$ and $\theta_j^1, \theta_j^2, \theta_j^3, \ldots$ converge. Let $\ell_j = \lim_{n\to\infty} \ell_j^n$ and $\theta_j = \lim_{n\to\infty} \theta_j^n$, $j = 1, 2, \ldots, 3g-3$.

We deform next the surface Σ in the following fashion. If $\ell_j = 0$ then we replace the decomposing curve α_j of the pants decomposition $\mathcal{P}$ by a node. Do that for each j with $\ell_j = 0$. That construction yields a stable surface Σ^* of genus g. The identity mapping $\Sigma \to \Sigma$ induces a strong deformation $f : \Sigma \to \Sigma^*$.

By the above mentioned properties 1 and 2 of the representative (Σ, Y_n, τ) we deduce that τ induces an orientation reversing involution $\tau : \Sigma^* \to \Sigma^*$. Also the decomposition $\mathcal{P}$ of Σ into pairs of pants gives a similar decomposition $\mathcal{P}$ of Σ^* into pairs of pants.

On Σ^* define a complex structure Y^* by the Fenchel–Nielsen coordinates ℓ_j and θ_j with respect to the decomposition $\mathcal{P}$ of Σ^* into pairs of pants. That complex structure Y^* is uniquely defined up to Dehn twists around those decomposing curves α_j of $\mathcal{P}$ that are not nodes and up to deformations by mappings homotopic to the identity mapping in each pair in each pair of pants belonging to $\mathcal{P}$.

The diagram

$$\begin{array}{ccc} (\Sigma, Y_n) & \xrightarrow{f} & (\Sigma^*, Y^*) \\ \downarrow{\scriptstyle\tau} & & \downarrow{\scriptstyle\tau} \\ (\Sigma, Y_n) & \xrightarrow{f} & (\Sigma^*, Y^*) \end{array} \tag{5.9}$$

commutes.

Here $f : (\Sigma, Y_n) \to (\Sigma^*, Y^*)$ is a strong deformation and each mapping $\tau : (\Sigma, Y_n) \to (\Sigma, Y_n)$ is an antiholomorphic involution mapping $\mathcal{P}$ onto itself.

Consider next the decomposition $\mathcal{P}$ on Σ into pairs of pants as *an oriented decomposition.* That gives an orientation also for the decomposition $\mathcal{P}$ of Σ^* into pairs of pants. The mapping τ maps each pair of pants belonging to $\mathcal{P}$ onto some other (or possibly the same) pair of pants in $\mathcal{P}$. Therefore, as oriented decompositions of Σ into pairs of pants $\tau(\mathcal{P})$ and $\mathcal{P}$ are different.

Since each mapping $\tau : (\Sigma, Y_n) \to (\Sigma, Y_n)$ is antiholomorphic, the Fenchel-Nielsen coordinates of Y_n with respect to the oriented pants decomposition $\tau(\mathcal{P})$ are obtained by a permutation of the coordinates ℓ_j^n and the coordinates θ_j^n and by possibly replacing some of the coordinates θ_j^n by the coordinates $2\pi - \theta_j^n$. Observe that the coordinates of Y_n with respect to $\tau(\mathcal{P})$ are always obtained from the coordinates of Y_n with respect to $\mathcal{P}$ by the same continuous transformation which does not depend on n.

We conclude, by continuity, that also the coordinates of Y^* with respect to the oriented pants decomposition $\tau(\mathcal{P})$ are obtained by the same transformation from the coordinates of Y^* with respect to the pants decomposition $\mathcal{P}$. Then we conclude that by deforming the complex structure Y^* by additional Dehn twists around some of the decomposing curves α_j that are not nodes of Σ^*, and by deforming Y^* in each pair of pants of $\mathcal{P}$ by a mapping homotopic to the identity mapping, we may actually suppose that the mapping $\tau : (\sigma^*, Y^*) \to (\Sigma^*, Y^*)$ is an antiholomorphic involution. Details here are easy but tedious. Therefore (Σ^*, Y^*) defines a point in $\overline{M_{\mathbf{R}}^g}$.

From the construction it follows then immediately that

$$[(\Sigma, X_n)] \to [(\Sigma^*, Y^*)] \quad \text{as} \quad n \to \infty$$

in $\overline{M}^g_{\mathbf{R}}$. This proves the theorem.

By Theorem 5.9.1, $\overline{M}^g_{\mathbf{R}}$ is the closure of $M^g_{\mathbf{R}}$. The moduli space $M^g_{\mathbf{R}}$ is, on the other hand, the union of finitely many components $V(g, n, k)$. Since, by Theorem 5.10.2, each one of them has compact closure in $\overline{M}^g_{\mathbf{R}}$, also the closure of $M^g_{\mathbf{R}}$ in $\overline{M}^g_{\mathbf{R}}$ is compact. That remark proves the main result of this Section:

Theorem 5.10.3 *The real moduli space of symmetric genus g Riemann surfaces, $\overline{M}^g_{\mathbf{R}}$, is compact for $g > 1$.*□

Observe that in the above we have considered only the case of hyperbolic Riemann surfaces, i.e., we have assumed everywhere that $g > 1$. The results hold also in the case of genus 1 or 0 Riemann surfaces but have to be shown by the methods of Section 5.3. The case of symmetric genus 0 Riemann surfaces is trivial, since, in that case, the moduli space reduces to a set of two points.

5.11 Review on results concerning the analytic structure of moduli spaces of compact Riemann surfaces

The focus of this monograph has been on that part of the theory of Teichmüller spaces that can be derived studying multipliers of Möbius transformations. That leads to parametrizations of Teichmüller spaces by geodesic length functions. These are real analytic parameters, while the Teichmüller space of genus g compact Riemann surfaces is actually a $3g - 3$ dimensional complex manifold. Teichmüller spaces of *non–classical* surfaces are only real analytic (and not complex) manifolds, but the complex structure of the classical Teichmüller spaces plays an important rôle in this non–classical theory. For that reason we provide, in this Section, a review of this complex theory. Everything here will be presented without proofs, but exact references will be given.

During recent years three excellent monographs, which treat the complex analytic theory of Teichmüller spaces, have appeared. They were written by Fredrick P. Gardiner ([32]), Olli Lehto ([60]) and by Subhasis Nag ([69]). Especially the above cited monograph of Gardiner and that of Nag give a good introduction the modern parts of this complex theory.

On page 148 we have described the complex structure of the Teichmüller space of compact smooth genus g Riemann surface, $g > 1$. Main result concerning the complex structure is Theorem 4.4.2 (on page 148), which states that the Teichmüller space T^g of compact genus g, $g > 1$, Riemann surfaces is a complex manifold of complex dimension $3g-3$. By the constructions related to the complex structure it is easy to see that elements of the modular group are holomorphic automorphisms of the Teichmüller space of compact genus g Riemann surfaces. Royden has shown (in [71] and in [72]), furthermore, that the modular group is, for $g > 2$, the *full* group of holomorphic automorphisms of T^g. Using this result one can draw some conclusions concerning the structure of the moduli space.

To that end we need the following theorem of Henri Cartan [24, Théorème 1]:

Theorem 5.11.1 *Let X be a complex manifold and G a properly discontinuous group of holomorphic automorphisms of X. The quotient X/G is a normal complex space.*

The normality is a condition concerning the singularities of the moduli space. This result simply means that the singularities are not too bad.

By Theorem 4.10.5 (on page 163), the modular group Γ^g acts properly discontinuously on the Teichmüller space T^g. Since elements of the modular group are holomorphic automorphisms of T^g, Theorem 5.11.1 implies:

Theorem 5.11.2 *The moduli space M^g, $g > 1$, of smooth compact genus g Riemann surfaces is a normal complex space.*

By means of the Geometric Invariant Theory algebraic geometers, David Mumford, Walter L. Baily, David Gieseker, Finn Knudsen and others, have shown a stronger result ([50][51],[52],[65], [66],[67], [68], [38]):

Theorem 5.11.3 *The moduli space, M^g, of smooth compact genus g Riemann surfaces, $g > 1$, is a quasiprojective algebraic variety. Its compactification $\overline{M^g}$ is a projective algebraic variety.*

This is one of the most important results in the theory of Riemann surfaces and algebraic curves.

After the success of the Geometric Invariant Theory in compactifying the moduli space and studying its structure a natural problem was to prove the same results using Teichmüller theory. That this is possible was announced as early as in 1973 ([11]). The details turned out to be difficult. Concerning the compactification it was, for many years, possible to prove, using exclusively Teichmüller theory, only that the moduli space of stable

Riemann surfaces of a given genus is a compact Hausdorff space (Theorem 5.11.2 above). Ideas to this proof are due to Lipman Bers and appeared already in [11]. For another account of these facts see also the notes of Joe Harris ([37]).

To prove, using arguments of analytic geometry, that the moduli space of stable Riemann surfaces is a projective variety, took much longer. First geometric proof for this fact is due to Scott Wolpert ([101]). Recently Frank Herrlich ([41]) has been able to complete the original arguments of Bers ([11]) completing this line of investigations.

In view of the rich complex analytic theory of Teichmüller spaces and moduli spaces of classical Riemann surfaces, it is natural to ask what can be said about the possible analytic structure of the Teichmüller spaces and the moduli spaces of non–classical Riemann surfaces.

Let Σ be a non–classical topological surface and Σ^c its complex double. Then Σ^c has an orientation reversing involution $\sigma : \Sigma^c \to \Sigma^c$ such that $\Sigma = \Sigma^c/\langle\sigma\rangle$. Recall that by Theorem 4.5.1 (on page 149) we may identify the Teichmüller space $T(\Sigma)$ of the non–classical compact surface Σ with the fixed–point set $T(\Sigma^c)_{\sigma^*}$ of the induced involution $\sigma^* : T(\Sigma^c) \to T(\Sigma^c)$. This involution is antiholomorphic. Locally the involution σ^* can be modelled as the complex conjugation in $\mathbf{C}^{3g-3}$. Since the fixed–point set of such an antiholomorphic involution is always a real analytic manifold we have:

Theorem 5.11.4 *The Teichmüller space of a non–classical compact surface Σ of genus g, $g > 1$, is a real analytic manifold of real dimension $3g-3$*

Next question is what can be said about the moduli space $M^g_{\mathbf{R}}$ of non–classical Riemann surfaces of genus g. We consider here only the general case $g > 1$.

In [79, Theorems 2.1 and 9.3] (see also [75]) Mika Seppälä and Robert Silhol have shown the following result:

Theorem 5.11.5 *Components of the moduli space $M^g_{\mathbf{R}}$ of non–classical Riemann surfaces of genus g, $g > 1$, are semi–algebraic varieties and irreducible real analytic spaces.*

Let $\pi : \overline{M^g_{\mathbf{R}}} \to \overline{M^g}$ be the mapping that forgets the real structure. We observed above that the moduli space $\overline{M^g}$ is a complex projective variety and a normal complex space. Another natural problem is to study the structure of $\pi(\overline{M^g_{\mathbf{R}}})$ in $\overline{M^g}$.

To that end, recall that the moduli space $\overline{M^g}$ of genus g compact stable Riemann surfaces can be viewed as the moduli space of stable complex algebraic curves. Likewise, $M^g_{\mathbf{R}}$ is the moduli space of stable real algebraic curves. Let $\tau : \overline{M^g} \to \overline{M^g}$ be the mapping that takes the isomorphism class of a complex algebraic curve onto that of its complex conjugate. Using techniques described in [67] it is possible to embedd $\overline{M}^g$ into a complex projective space $\mathbf{P}^N(\mathbf{C})$ in such a way that the diagram

$$\begin{array}{ccccc} \overline{M}^g_{\mathbf{R}} & \to & \overline{M}^g & \hookrightarrow & \mathbf{P}^N(\mathbf{C}) \\ \downarrow{\scriptstyle Id} & & \downarrow{\scriptstyle \tau} & & \downarrow{\text{compl. conj.}} \\ \overline{M}^g_{\mathbf{R}} & \to & \overline{M}^g & \hookrightarrow & \mathbf{P}^N(\mathbf{C}) \end{array} \tag{5.10}$$

is commutative. This means, in particular, that

$$\pi(\overline{M}^g_{\mathbf{R}}) \subset \left(\overline{M}^g\right)_\tau = \overline{M}^g(\mathbf{R})$$

where $\left(\overline{M}^g\right)_\tau$ denotes the fixed-point set of the involution $\tau : \overline{M}^g \to \overline{M}^g$, $\overline{M}^g(\mathbf{R})$ is the set of real points of $\overline{M}^g$ and we have interpreted $\overline{M}^g$ as a subset of $\mathbf{P}^N(\mathbf{C})$.

Regarding diagram (5.10) Clifford Earle showed in [26] that $\overline{M}^g_{\mathbf{R}} \neq \overline{M}^g(\mathbf{R})$. Another proof and a direct construction for this fact has been given by Goro Shimura in [84]. Earle's proof uses Teichmüller theory while Shimura gives an explicit family of polynomials that define complex algebraic curves, with real moduli, that are not real curves. This observation has been a motivation in the study of the properties of the moduli space $\overline{M}^g_{\mathbf{R}}$.

In [78] Mika Seppälä has shown:

Theorem 5.11.6 *Assume that $g > 3$. The image $\pi(\overline{M}^g_{\mathbf{R}})$ of $\overline{M}^g_{\mathbf{R}}$ in $\overline{M}^g$ is a connected semialgebraic variety and the quasi-regular real part of $\overline{M}^g$.*

Recall that the *quasi-regular real part*, as defined by Aldo Andreotti and Per Holm in [9], consists of those points of $\overline{M}^g(\mathbf{R})$ where the local dimension of $\overline{M}^g(\mathbf{R})$ is maximal.

This result has recently been extended to the case $g = 3$ by Robert Silhol (see [88]). In that pepr Silhol has also shown that the result *cannot* be extended ot the case $g = 2$. Silhol has also extensively studied moduli problems of real abelian varieties and obtained important results for them.

Observe that Theorem 5.10.2 implies that $\pi(\overline{M}^g_{\mathbf{R}})$ is a compact subset of $\overline{M}^g$. We can characterize $\pi(\overline{M}^g_{\mathbf{R}})$ as the subset of $\overline{M}^g$ consisting of isomorphism classes of those Riemann surfaces X on which $\mathbf{Z}_2 = \mathbf{Z}/2\mathbf{Z}$ acts *via an antiholomorphic involution.*

This formulation leads to the following generalization. Let G be a finite group. Set

$$\overline{M}^g_G = \{[X] \mid G \subset \operatorname{Isom} X\}. \tag{5.11}$$

Here $\operatorname{Isom} X$ denotes the isometry group of X. Theorem 4.1 of my other lecture in this volume means that it is not possible to use the construction of Sections 1 – 3 to study this set $\overline{M}^g_G$ in the general case.

In the general case $\overline{M}^g_G$ is *not* compact. It is an interesting problem to characterize the closure of $\overline{M}^g_G$ in $\overline{M}^g$.

We conclude this Section by the following remarks. The modern generalizations of these arguments to moduli problems of real algebraic geometry were started by Clifford Earle ([26]) and by Goro Shimura ([84]). Earlier these moduli problems were extensively studied by Felix Klein (cf. e.g. [46]). These investigations were continued independently by S. Natanzon ([70]), Mika Seppälä (see e.g. [77] and [78]), Robert Silhol (see e.g. [88]) and jointly by Seppälä and Silhol ([79]).

Appendix A

Hyperbolic metric and Möbius groups

A.1 Length and area elements

The Möbius transformation

$$g(z) = \frac{az+b}{\overline{b}z+\overline{a}} \tag{A.1}$$

with $|a|^2 - |b|^2 = 1$ maps the unit disk $D = \{z \in \mathbf{C} \mid |z| < 1\}$ conformally onto itself. For $z_1, z_2 \in D$ let $w_j = g(z_j)$, $j = 1, 2$. From (A.1) we get

$$w_1 - w_2 = \frac{z_1 - z_2}{(\overline{b}z_1 + \overline{a})(\overline{b}z_2 + \overline{a})}$$

and

$$1 - \overline{w}_1 w_2 = \frac{1 - \overline{z}_1 z_2}{(b\overline{z}_1 + a)(\overline{b}z_2 + \overline{a})}.$$

Hence

$$\left|\frac{z_1 - z_2}{1 - \overline{z}_1 z_2}\right| = \left|\frac{w_1 - w_2}{1 - \overline{w}_1 w_2}\right|. \tag{A.2}$$

Letting z_1 approach z_2 (A.2) becomes

$$\frac{|dz|}{1 - |z|^2} = \frac{|dw|}{1 - |w|^2}.$$

This shows that the Riemannian metric whose element of length is

$$d_D s = ds = \frac{2|dz|}{1 - |z|^2} \tag{A.3}$$

is invariant under conformal self–mappings of the unit disk. In this metric every rectifiable arc γ has the length

$$\int_\gamma \frac{2|dz|}{1-|z|^2},$$

and every measurable set E has the area

$$\iint_E \frac{4dxdy}{(1-|z|^2)^2}.$$

The metric defined by the line element (A.3) is called *the hyperbolic metric* of D. In this metric the shortest arc from 0 to any other point is along the radius. Hence geodesics are euclidean circles orthogonal to $\partial D = \{z \in \mathbf{C} \mid |z| = 1\}$. Such geodesics are also called *h–lines* or *hyperbolic lines.* D together with the hyperbolic metric is called *the hyperbolic plane.*

Let $\sigma(z) = 2/(1-|z|^2)$. Then the Gaussian curvature of the hyperbolic metric is

$$K = -\frac{2}{\sigma(z)^2}\frac{\partial^2 \log(\sigma(z)^2)}{\partial z \partial \bar{z}} = -(1-|z|^2)^2 \frac{1-|z|^2+z\bar{z}}{(1-|z|^2)^2} = -1. \tag{A.4}$$

The constant 2 was added in (A.3) in order to get a metric of constant curvature -1. Generally speaking any Riemannian metric of curvature -1 is called *hyperbolic.*

The hyperbolic geometry can also be carried over to the upper half–plane $U = \{z \in \mathbf{C} \mid \operatorname{Im} z > 0\}$. The element of length that corresponds to (A.3) is

$$d_U s = \frac{|dz|}{\operatorname{Im} z}. \tag{A.5}$$

The upper half–plane together with this metric is another model for the hyperbolic plane. Sometimes it is more convenient to work in U instead of D. Therefore we frequently switch back and forth between these two hyperbolic planes.

The hyperbolic distance from 0 to $r > 0$ is

$$\int_0^r \frac{2dr}{1-r^2} = \log\frac{1+r}{1-r}.$$

For the hyperbolic distance $d_D(z_1, z_2)$ of arbitrary points $z_1, z_2 \in D$ we have the formula

$$d_D(z_1, z_2) = \log\frac{1+t}{1-t}, \tag{A.6}$$

where $t = |z_1 - z_2|/|1 - \bar{z}_1 z_2|$. A similar formula with

$$t = \frac{|z_1 - z_2|}{|\overline{z_1} - z_2|}$$

holds for the hyperbolic metric of the upper half–plane.

A.2 Isometries of the hyperbolic metric

By (A.2) all Möbius–transformations mapping the unit disk onto itself are isometries of the hyperbolic metric. Conversely, let $f : D \to D$ be an isometry of the hyperbolic metric. We want to show that f is either a Möbius–transformation of the complex conjugate of a Möbius–transformation.

To that end consider, in the stead of f, the mapping

$$F(z) = \frac{(1 - \overline{f(0)}f(1/2))}{(f(1/2) - f(0))} \frac{(f(z) - f(0)))}{(1 - \overline{f(0)}f(z))}.$$

Then F is an isometry as well and it maps the interval $[0, 1]$ onto itself keeping the origin fixed. It suffices to show that F is either a Möbius–transformation or the complex conjugate of a Möbius–transformation.

Let C_r denote the euclidean circle $\{z \in \mathbf{C} \mid |z| = r\}$ for $0 \leq r \leq 1$. Observe that C_r is a circle in the hyperbolic metric as well. Since F is an isometry of the hyperbolic metric and since $F(0) = 0$, $F(C_r) = C_r$ for each r.

Let $re^{i\varphi}$ be an arbitrary point of C_r. Then F maps the arc of C_r with end–points r and $z = re^{i\varphi}$ onto some arc with end–points r and $re^{i\varphi_{f(z)}}$.

F being an isometry of the hyperbolic metric these two arcs must have the same hyperbolic length. We conclude that either $\varphi_{f(z)}=\varphi$ or $\varphi_{f(z)} = -\varphi$ for each $z = re^{i\varphi}$. Since F is continuous we conclude then that either $\varphi_{f(z)} = \varphi$ or $\varphi_{f(z)} = -\varphi$ for all $z = re^{i\varphi} \in D$.

This proves the following result:

Theorem A.2.1 *The group of isometric self–mappings of the hyperbolic unit disk D is the group of conformal or anticonformal self–mappings of D.*

A.3 Geometry of the hyperbolic metric

A ray from a point z in the hyperbolic plane is an infinite arc on a h–line with an end–point z. Angles in the hyperbolic plane can be defined just like angles in the Euclidean geometry; they are bounded by two hyperbolic rays starting from one point. Angles can be measured just like angles in the Euclidean geometry. It follows that hyperbolic angles are euclidean angles.

The following properties of the hyperbolic metric follow easily from the above:

- Hyperbolic circles are Euclidean circles but their centers do not usually coincide.
- There is a unique h–line through any two distinct points of the hyperbolic plane.

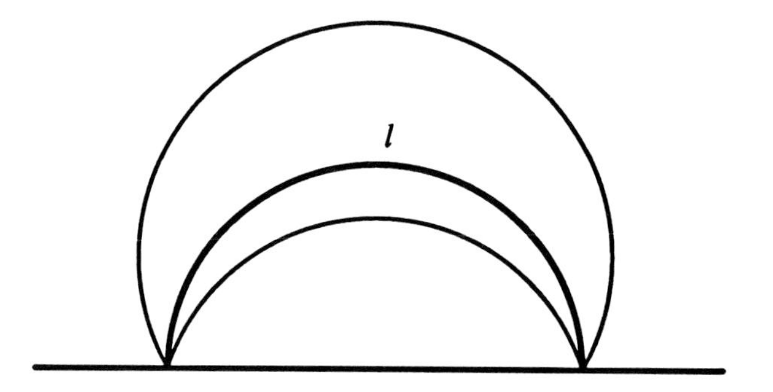

Figure A.1: Points whose distance from l is at most r.

- Two distinct h–lines intersect in at most one point of the hyperbolic plane.
- Given any two h–lines l_1 and l_2, there is an isometry g such that $g(l_1) = l_2$.
- Given any h–line l and any point z, there is a unique h–line through z and orthogonal to l.

The distance of a point z from a h–line l is measured along the unique h–line through z perpendicular to l. The imaginary axes l is a h–line in the upper half–plane. Computing by (A.6) we conclude that the set of points of U which lie at a given distance r from l consists of two euclidean rays starting from the origin and forming equal angles with l. Let φ be that angle. One computes further that $r \to \infty$ as $\varphi \to \pi/2$.

Let $d_U(z, l)$ denote the distance of the point z from the h–line l. We conclude that for an arbitrary h–line l in U the set

$$\{z \in U \mid d_U(z, l) < r\}$$

is the crescent in figure (A.1).

Let D_r be the disk of hyperbolic radius r with center at the origin of the hyperbolic unit disk D. Then D_r is also an euclidean disk. Its euclidean radius R equals $R = \operatorname{arctanh}(r/2)$ by (A.6). Straightforward integration then gives the following result.

Theorem A.3.1 *The hyperbolic area of a hyperbolic disk of radius r is $2\pi(\cosh r - 1)$. The hyperbolic length of a hyperbolic circle of radius r is $2\pi \sinh r$.*

Observe that the area of the hyperbolic disk D_r can also be expressed as $4\pi \sinh^2(r/2)$.

A.4 Matrix groups

Denote by $M(n,\mathbf{R})$ the real vector space of real $n \times n$ matrices

$$A = \begin{pmatrix} a_{11} & a_{12} & \cdots & a_{1n} \\ a_{21} & a_{22} & \cdots & a_{2n} \\ \vdots & \vdots & \ddots & \vdots \\ a_{n1} & a_{n2} & \cdots & a_{nn} \end{pmatrix}.$$

The linear mapping $\mathbf{R}^n \to \mathbf{R}^n$ defined by A is bijective if and only if $\det A \neq 0$. In this case A is *non-singular* and the inverse matrix A^{-1} satisfies

$$AA^{-1} = A^{-1}A = I = (\delta_{ij})$$

where δ_{ij} is the Kronecker symbol. Denote by

$$\mathrm{GL}(n,\mathbf{R}) \quad \text{(“General Linear Group”)}$$

the group of real non-singular $n \times n$-matrices.

The complex vector space of complex $n \times n$-matrices $A = (a_{ij})$ is denoted by $M(n,\mathbf{C})$. Also in this case A is non-singular if and only if $\det A \neq 0$. Let $\mathrm{GL}(n,\mathbf{C})$ denote the group of complex non-singular $n \times n$-matrices. In a natural way, $\mathrm{GL}(n,\mathbf{R})$ is a subgroup of $\mathrm{GL}(n,\mathbf{C})$.

If A, $B \in M(n,\mathbf{C})$, then

$$\det(AB) = \det A \det B = \det(BA).$$

If $A \in \mathrm{GL}(n,\mathbf{C})$, then $\det A \det A^{-1} = 1$ and $\det A^{-1} = (\det A)^{-1}$. Especially,

$$\det(ABA^{-1}) = \det(AA^{-1}B) = \det B.$$

It follows that matrices $A \in \mathrm{GL}(n,\mathbf{C})$ with $\det A = 1$ constitute a group. This subgroup of $\mathrm{GL}(n,\mathbf{C})$ is denoted by

$$\mathrm{SL}(n,\mathbf{C}) \quad \text{(“Special Linear Group”).}$$

Similarly, $\mathrm{SL}(n,\mathbf{R}) = \{A \in \mathrm{GL}(n,\mathbf{R}) \mid \det A = 1\}$ is a subgroup of $\mathrm{GL}(n,\mathbf{R})$. Moreover, $\mathrm{SL}(n,\mathbf{R}) \subset \mathrm{SL}(n,\mathbf{C})$.

Example. Let $A = \left(\begin{smallmatrix} a & b \\ c & d \end{smallmatrix}\right) \in \mathrm{GL}(2,\mathbf{C})$. Then we have $\det A = ad - bc \neq 0$. If $A^{-1} = \left(\begin{smallmatrix} \alpha & \beta \\ \gamma & \delta \end{smallmatrix}\right)$, then

$$\begin{pmatrix} 1 & 0 \\ 0 & 1 \end{pmatrix} = \begin{pmatrix} a & b \\ c & d \end{pmatrix}\begin{pmatrix} \alpha & \beta \\ \gamma & \delta \end{pmatrix} = \begin{pmatrix} a\alpha + b\gamma & a\beta + b\delta \\ c\alpha + d\gamma & c\beta + d\delta \end{pmatrix},$$

and hence

$$\alpha = \frac{d}{ad-bc}, \ \beta = \frac{-b}{ad-bc}, \ \gamma = \frac{-c}{ad-bc}, \ \delta = \frac{a}{ad-bc}. \ \square$$

Let $I = \begin{pmatrix} 1 & 0 \\ 0 & 1 \end{pmatrix} \in \mathrm{SL}(2,\mathbf{C})$ and $-I = \begin{pmatrix} -1 & 0 \\ 0 & -1 \end{pmatrix}$. Then $A(-I) = (-I)A = -A$ for all $A \in M(2,\mathbf{C})$. Denote

$$N = \{I, -I\}.$$

Since $I^2 = I$, $(-I)^2 = I$ and $I(-I) = (-I)I = -I$, N is a subgroup of $\mathrm{SL}(2,\mathbf{C})$. Moreover, since $AIA^{-1} = I$ and $A(-I)A^{-1} = -I$ for all $A \in \mathrm{SL}(2,\mathbf{C})$, N is a normal subgroup of $\mathrm{SL}(2,\mathbf{C})$ and of $\mathrm{SL}(2,\mathbf{R})$. The quotient groups

$$\begin{aligned} \mathrm{SL}(2,\mathbf{C})/N &= \mathrm{PSL}(2,\mathbf{C}), \quad (\text{"Projective Special Linear Group"}) \\ \mathrm{SL}(2,\mathbf{R})/N &= \mathrm{PSL}(2,\mathbf{R}) \end{aligned}$$

consist of equivalence classes $(A, -A)$, $A \in \mathrm{SL}(2,\mathbf{C})$.

The vector space $M(n,\mathbf{C})$ can be identified with the euclidean space $\mathbf{C}^{n\times n}$. Hence $M(n,\mathbf{C})$ has a natural topology. The ε–neighbourhood of the matrix $A = (a_{ij})$ consists of the matrices $B = (b_{ij})$ for which

$$\|A - B\| = \left(\sum |a_{ij} - b_{ij}|^2\right)^{1/2} < \varepsilon.$$

The same topology is obtained if ε–neighbourhoods are defined by the metric

$$d(A,B) = \max_{i,j} |a_{ij} - b_{ij}|.$$

If $A_k = (a_{ij}(k))$ and $A = (a_{ij})$, then

$$\lim_{k\to\infty} A_k = A \iff \lim_{k\to\infty} a_{ij}(k) = a_{ij} \ \text{ for all } i \text{ and } j.$$

Hence the mapping $\det : M(n,\mathbf{C}) \to \mathbf{C}$, $A \mapsto \det A$, is continuous. For $A = (a_{ij}) \in M(n,\mathbf{C})$, let

$$\mathrm{tr}A = a_{11} + a_{22} + \cdots + a_{nn}.$$

Then also the mapping $\mathrm{tr} : M(n,\mathbf{C}) \to \mathbf{C}$ is continuous.

We have seen that $\mathrm{GL}(n,\mathbf{R})$, $\mathrm{SL}(n,\mathbf{R})$, $\mathrm{GL}(n,\mathbf{C})$ and $\mathrm{SL}(n,\mathbf{C})$ are *both* groups *and* topological spaces. Moreover,

$$A_n^{-1} \to A^{-1} \text{ and } A_nB_n \to AB$$

whenever $A_n \to A$ and $B_n \to B$.

Generally, let G be an arbitrary group. Suppose that G is also a topological space. If the mappings

$$\begin{array}{rcl} x & \mapsto & x^{-1} \quad (G \to G) \\ (x,y) & \mapsto & xy \quad (G \times G \to G) \end{array}$$

are continuous, then G is a *topological group.* Suppose that every point $x \in G$ has a countable base of neighbourhoods. Then the mappings $x \mapsto x^{-1}$ and $(x,y) \mapsto xy$ are continuous if $x_n^{-1} \to x^{-1}$ and $x_n y_n \to xy$ whenever $x_n \to x$ and $y_n \to y$. Hence all subgroups of $\mathrm{GL}(n, \mathbf{C})$ and $\mathrm{SL}(n, \mathbf{C})$ are topological groups.

For any $y \in G$, the space $G \times \{y\}$ has a natural topology whose open sets are of the form $A \times \{y\}$, $A \subset G$ open. Then the mapping $x \mapsto (x,y)$ is a homeomorphism $G \to G \times \{y\}$, and the mapping $(x,y) \mapsto xy$ is a continuous surjection $G \times \{y\} \to G$. The composition

$$x \mapsto (x,y) \mapsto xy$$

is a continuous surjection $G \to G$ whose inverse $x \mapsto xy^{-1}$ is continuous. Hence we have

Theorem A.4.1 *Let G be a topological group. Then for every $y \in S$, the mappings $x \mapsto xy$ and $x \mapsto yx$ are homeomorphisms $G \to G$.* □

Definition A.4.1 *A topological group G is* discrete *if the topology of G is discrete, i.e., if all subsets of G are open.*

Theorem A.4.2 *Let G be a topological group. Suppose there exists $g \in G$ such that $\{g\}$ is open. Then G is discrete.*

Proof. Let $y \in G$. Since the mapping $x \mapsto x(g^{-1}y)$ is homeomorphic, the set $\{g(g^{-1}y)\} = \{y\}$ is open. □

Let $H \subset G$ be a normal subgroup. We state without a proof that the quotient group G/H becomes a topological group if it is equipped with the topology co–induced by the projection map $G \to G/H$.

In the following, we shall consider subgroups of $M(2, \mathbf{C})$. Let

$$P : \mathrm{SL}(2, \mathbf{C}) \to \mathrm{PSL}(2, \mathbf{C})$$

by the projection for which $P(A) = P(-A)$. Then P is locally injective. If $\mathrm{PSL}(2, \mathbf{C})$ is equipped with the topology co–induced by P and if $\mathbf{A} \in \mathrm{PSL}(2, \mathbf{C})$ and $P^{-1}(\mathbf{A}) = \{A_1, A_2\}$, then $\mathbf{A}$ has a neighbourhood U such that $P^{-1}(U)$ consists of disjoint neighbourhoods of A_1 and A_2. Moreover, PSL(2, **C**) and SL(2, **C**) and all their subgroups are topological groups.

By the above definition, a topological group G is discrete if G contains no accumulation points, i.e., if the conditions

$$A_n \to A; \quad A_n, A \in G$$

always imply that there exists an n_0 such that $A_n = A$ for all $n \geq n_0$. We show next that a discrete group $G \subset \mathrm{SL}(2, \mathbf{C})$ has in fact no accumulation points in SL(2, **C**).

Theorem A.4.3 *Let $G \subset \mathrm{SL}(2, \mathbf{C})$ be a group. Then G is discrete if and only if the conditions*

$$A_n \to A \text{ and } A_n \in G,\ A \in M(2, \mathbf{C})$$

always imply that there exists an n_0 such that $A_n = A$ for all $n \geq n_0$.

Proof. Suppose that G is discrete. Let $\{A_1, A_2, \ldots\} \subset G$ and $A \in M(2, \mathbf{C})$ such that $A_n \to A$. Since

$$1 = \det A_n \to \det A,$$

we have $\det A = 1$, i.e., $A \in \mathrm{SL}(2, \mathbf{C})$. Since SL(2, **C**) is a topological group,

$$A_n^{-1} \to A^{-1} \text{ and } T_n = A_n^{-1} A_{n+1} \to A^{-1} A = I.$$

Since $T_n \in G$, there exists an n_0 such that

$$n \geq n_0 \Rightarrow T_n = I.$$

Then $A_{n_0} = A_{n_0+1} = A_{n_0+2} = \ldots$. Since $A_n \to A$, we have $A_n = A$ for all $n \geq n_0$.

Conversely, suppose that the conditions of the theorem hold. Let $\{A_1, A_2, \ldots, A\} \subset G$ such that $A_n \to A$. Then there exist an n_0 such that $A_n = A$ for all $n \geq n_0$. Hence G is discrete. □

Let $G \subset \mathrm{PSL}(2, \mathbf{C})$ be a group and let $\mathbf{A}_n \in G$. Suppose that we can choose $A_n \in P^{-1}(\mathbf{A}_n)$ such that $A_n \to A \in M(2, \mathbf{C})$. Since $A_n \in \mathrm{SL}(2, \mathbf{C})$, we have $A \in \mathrm{SL}(2, \mathbf{C})$ by the preceding proof. Moreover,

$$A_n^{-1} \to A^{-1} \text{ and } T_n = A_n^{-1} A_{n+1} \to I.$$

Suppose that the matrices A_n are distinct. Then the sequence T_n contains infinitely many distinct elements since otherwise $T_n \to I$ would imply that $T_n = I$ and $A_n = A_{n+1}$ for $n \geq n_0$. Since $P(T_n) \in G$ and $P(T_n) \to P(I) \in G$, the group G would have $P(I)$ as a accumulation point. Then G would not be discrete.

Conversely, if no sequence $A_n \to A$ of the above type can be found, the group G is discrete. Hence Theorem A.4.3 has the following corollaries:

Corollary A.4.4 *A group $G \subset \mathrm{PSL}(2,\mathbf{C})$ is discrete if and only if the conditions*

$$A_n \to A, \quad P(A_n) \in G \text{ and } A \in M(2,\mathbf{C})$$

always imply that there exists an n_0 such that $A_n = A$ for all $n \geq n_0$. □

Corollary A.4.5 *Suppose that $G \subset \mathrm{PSL}(2,\mathbf{C})$ is not discrete. Then there exists an infinite sequence $A_n \in \mathrm{SL}(2,\mathbf{C})$, $n = 1, 2, \ldots$, of distinct elements such that $P(A_n) \in G$ and $A_n \to I$.* □

Consider discreteness of a group $G \subset \mathrm{SL}(2,\mathbf{C})$. By Corollary A.4.2, it suffices to find a point $A \in G$ such that the set $\{A\}$ is open. For instance, it suffices to show that $\{I\}$ is open, i.e.,

$$\inf\{\, \|X - I\| \mid X \in G,\, X \neq I \,\} > 0.$$

In other words, it suffices to show that the conditions $A_n \to I$, $A_n \in G$, imply that $A_n = I$ for all $n \geq n_0$.

Theorem A.4.6 *A group $G \subset \mathrm{SL}(2,\mathbf{C})$ is discrete if and only if the set*

$$\{\, A \in G \mid \|A\| \leq t \,\}$$

is finite for all $t > 0$.

Proof. If the set in question is finite for all $t > 0$, then G has no accumulation points and G is discrete.

Suppose conversely that the set $\{\, A \in G \mid \|A\| \leq t \,\}$ is infinite for some $t > 0$. Let $A_1, A_2, \ldots, \in G$ be distinct elements such that $\|A_n\| \leq t$. If $A_n = \begin{pmatrix} a_n & b_n \\ c_n & d_n \end{pmatrix}$, $a_n d_n - b_n c_n = 1$, then $|a_n| \leq t$, $|b_n| \leq t$, $|c_n| \leq t$ and $|d_n| \leq t$. Hence the sequence a_n has a convergent subsequence a_{n_i}, the sequence b_{n_i} has a convergent subsequence and so on. Therefore, we may assume that

$$\begin{aligned} a_n &\to \alpha, & b_n &\to \beta, \\ c_n &\to \gamma, & d_n &\to \delta. \end{aligned}$$

If $A = \begin{pmatrix} \alpha & \beta \\ \gamma & \delta \end{pmatrix}$, then $A_n \to A$. Since the matrices A_n are distinct, G is not discrete by Theorem A.4.3. □

Example. The *modular group* consists of all matrices $A = \begin{pmatrix} a & b \\ c & d \end{pmatrix} \in \mathrm{SL}(2,\mathbf{R})$ for which a, b, c, d are integers. By Theorem A.4.6 the modular group is discrete.

The Picard group consists of all matrices $A = \begin{pmatrix} a & b \\ c & d \end{pmatrix} \in \mathrm{SL}(2,\mathbf{C})$ for which a, b, c and d are complex integers, i.e., of the form $m + ni$ where m and n are integers. Also the Picard group is discrete.

A.5 Representation of groups

Let X be a non-empty set and let $F(X)$ be the set of all bijections $f : X \to X$. Then $F(X)$ is a group whose group operation is the composition of transformations. The transformation group $F(X)$ is said to *act* in X.

Let G be a subgroup of $F(X)$. A homomorphism $\varphi : G \to \mathrm{GL}(2, \mathbf{C})$ is a *representation* of G. If φ is injective, the representation is *faithful*. In this case we have an isomorphism

$$\varphi : G \to \varphi(G) \subset \mathrm{GL}(2, \mathbf{C}).$$

We shall also consider representations $\varphi : G \to \mathrm{SL}(2, \mathbf{C})$ and $\varphi : G \to \mathrm{PSL}(2, \mathbf{C})$.

Let $X = \hat{\mathbf{C}}$ be the extended complex plane and let G_1 be the group of all translations $S_\omega : z \mapsto z + \omega$, $\omega \in \mathbf{C}$. Then

$$S_\omega^{-1} = S_{-\omega} : z \mapsto z - \omega$$
$$S_{\omega_1} \circ S_{\omega_2} : z \mapsto z + \omega_1 + \omega_2.$$

If we interpret S_ω as a linear fractional transformation of the form $(az + b)/(cz + d)$, then

$$S_\omega(z) = z + \omega = \frac{1 \cdot z + \omega}{0 \cdot z + 1}.$$

Hence we obtain a mapping $\varphi : G_1 \to \mathrm{SL}(2, \mathbf{C})$ for which

$$\varphi(S_\omega) = \begin{pmatrix} 1 & \omega \\ 0 & 1 \end{pmatrix}.$$

Clearly, $\varphi(S_\omega^{-1}) = (\varphi(S_\omega))^{-1}$ and $\varphi(S_{\omega_1} \circ S_{\omega_2}) = \varphi(S_{\omega_1})\varphi(S_{\omega_2})$. Hence φ is a representation of G_1 in $\mathrm{SL}(2, \mathbf{C})$. Moreover, φ is faithful since

$$\varphi(S_\omega) = \begin{pmatrix} 1 & 0 \\ 0 & 1 \end{pmatrix} \Longleftrightarrow \omega = 0,$$

i.e., $\ker \varphi = \{\mathrm{id}\}$.

Secondly, let G_2 be the group of all stretchings $V_k : z \mapsto kz$, $k > 0$. Since

$$V_k(z) = kz = \frac{\sqrt{k}z + 0}{0 \cdot z + 1/\sqrt{k}},$$

we define now $\varphi : G_2 \to \mathrm{SL}(2, \mathbf{R})$ by setting

$$\varphi(V_k) = \begin{pmatrix} \sqrt{k} & 0 \\ 0 & 1/\sqrt{k} \end{pmatrix}.$$

Again, $\varphi(V_k^{-1}) = (\varphi(V_k))^{-1}$ and $\varphi(V_{k_1} \circ V_{k_2}) = \varphi(V_{k_1})\varphi(V_{k_2})$. It follows that φ is a faithful representation of G_2 in $\mathrm{SL}(2, \mathbf{R})$.

Thirdly, let G_3 be the group of all rotations

$$K_\vartheta : z \mapsto e^{i\vartheta} z.$$

Note that $K_{\vartheta_1} = K_{\vartheta_2}$ if and only if $\vartheta_1 \equiv \vartheta_2 \pmod{2\pi}$. Since

$$K_\vartheta(z) = \frac{e^{i\vartheta/2} z + 0}{0 \cdot z + e^{-i\vartheta/2}},$$

we define

$$\varphi(K_\vartheta) = \begin{pmatrix} e^{i\vartheta/2} & 0 \\ 0 & e^{-i\vartheta/2} \end{pmatrix}.$$

Then

$$\varphi(K_{\vartheta+2\pi}) = \begin{pmatrix} e^{i\vartheta/2} e^{i\pi} & 0 \\ 0 & e^{-i\vartheta/2} e^{-i\pi} \end{pmatrix} = \begin{pmatrix} -e^{i\vartheta/2} & 0 \\ 0 & -e^{-i\vartheta/2} \end{pmatrix} = -\varphi(K_\vartheta).$$

This shows that φ does not define a mapping $G_3 \to \mathrm{SL}(2, \mathbf{C})$ but a mapping $G_3 \to \mathrm{PSL}(2, \mathbf{C})$. Hence G_3 has a faithful representation in $\mathrm{PSL}(2, \mathbf{C})$.

Let $(F, \cdot)$ be a group. If S is a set of generators of F, then every $f \in F$ has a representation

$$f = x_1^{\varepsilon_1} x_2^{\varepsilon_2} \cdots x_n^{\varepsilon_n} \tag{A.7}$$

where $x_1, \ldots, x_n \in S$ and $\varepsilon_1, \ldots, \varepsilon_n = \pm 1$. An expression of the form (A.7) is called a *word* and the set S is called an *alphabet*. This representation of f as a word is not unique. By using normal rules of algebra, we get either the *empty word* or a representation of the form (A.7) such that $x_i^{\varepsilon_i} \neq x_{i+1}^{-\varepsilon_{i+1}}$, $i = 1, \ldots, n-1$. Even this *reduced representation* of f need not be unique. The neutral element e of F has always the empty word as a reduced representation. There exist groups whose neutral elements have no other reduced representations than the empty word.

Theorem A.5.1 *Every $f \in F$ has a unique reduced representation if and only if the neutral element of F has only the empty word as a reduced representation.* □

The group F is *free* if it has the property of Theorem A.5.1. More precisely, the definition can be formulated as follows.

Definition A.5.1 *Let S be a set of generators of a group F. If every mapping of S into an arbitrary (Abelian) group G can be extended to a homomorphism $F \to G$, then F is a* free (Abelian) group *and F is* generated freely *by S.*

Example. $F = (\{-1,1\}, \cdot)$ is a group whose neutral element is 1 and where -1 is its own inverse. The set $S = \{-1\}$ generates F. To show that F is not generated freely by S, let $G = (\mathbf{Z}, +)$ and define $\varphi : S \to G$ by $\varphi(-1) = -1$. Suppose that $\varphi : F \to G$ is a homomorphic extension. Then $\varphi(1) = 0$ since 0 is the neutral element of G. Hence

$$0 = \varphi(1) = \varphi((-1)\cdot(-1)) = \varphi(-1) + \varphi(-1) = -1 + (-1) = 2$$

which is impossible. Note that $(-1)^{+1}(-1)^{+1}$ is a reduced representation of 1. □

A proof of the following theorem can be found in standard textbooks of algebra (see [61] or [56]).

Theorem A.5.2 (Nielsen–Schreier) *Subgroups of a free group are free groups.* □

Theorem A.5.3 (Existence) *Let S be a non-empty set. Then there exists a free group F generated by S.*

Proof. Consider the Cartesian product

$$T = S \times \{-1, 1\}.$$

For any $a \in S$, denote

$$a^1 = (a, 1) \quad \text{and} \quad a^{-1} = (a, -1).$$

Choose the set T as an alphabet and let E be the set of all finite non–empty words. Let F contain the empty word e and all reduced words of E.

We define in F an operation as follows. Let u, $v \in F$. If $u = e$, let $uv = v$. Similarly, if $v = e$, let $uv = u$. In all other cases, u and v are reduced words of E. Then $uv \in E$. By cancelling from uv all pairs of the form a^1a^{-1} and $a^{-1}a^1$ we get either e or a uniquely defined $w \in F$. Define in the first case $uv = e$ and in the second case $uv = w$.

It follows that F is a group with e as the neutral element.

Let us identify a^1 with a. To show that F is generated freely by S, let G be an arbitrary group and $g : S \to G$ a mapping. Define $h : F \to G$ as follows: Let $h(e)$ be the neutral element of G. If $w \neq e$, then $w \in E$ and w has a unique reduced representation

$$w = a_1^{\varepsilon_1} a_2^{\varepsilon_2} \dots a_n^{\varepsilon_n}$$

where $\varepsilon_i = \pm 1$, $a_i \in S$, $i = 1, \dots, n$. Then we set

$$h(w) = [g(a_1)]^{\varepsilon_1} [g(a_2)]^{\varepsilon_2} \dots [g(a_n)]^{\varepsilon_n}.$$

It follows that $h : F \to G$ is a homomorphism and $h|S = g$. □

Theorem A.5.4 *Every group is isomorphic to a quotient group of a free group.*

Proof. For a group G, choose a set S of generators of G. Let F be the free group generated by S. Then the inclusion map $g : S \to G$ has a homomorphic extension $h : F \to G$. Since $S = g(S) \subset h(F)$ and S generates G, we have $h(F) = G$. Let $K = \ker h$ be the kernel of h. then G and F/K are isomorphic. □

In the proof of Theorem A.5.4, K is a normal subgroup of the free group F. Let R be a set of generators of K. (By the Nielsen–Schreier Theorem, we may suppose that R generates K freely.)

The set S determines F and the set R determines K. Hence the group

$$G \simeq F/K$$

is determined by the sets S and R. The elements of S are called *generators* of G and the elements of R are called *defining relations* of G.

Let $w \in R$. If $K \neq \{e\}$, we may suppose that $w \neq e$. Then w is a reduced word

$$w = a_1^{\varepsilon_1} a_2^{\varepsilon_2} \cdots a_n^{\varepsilon_n}$$

where $a_i \in S$ and $\varepsilon_i = \pm 1$, $i = 1, \ldots, n$. Since $w \in K$, w represents the neutral element e of G. This is denoted by the equation

$$a_1^{\varepsilon_1} a_2^{\varepsilon_2} \cdots a_n^{\varepsilon_n} = e$$

which is a traditional form of a defining relation.

Example. Let $g : z \mapsto e^{i\frac{2\pi}{n}} z$. Then the mappings $g^0 = \mathrm{id}$, $g^1 = g$, $g^2 = g \circ g$, $g^3 = g \circ g \circ g$, ..., g^{n-1} are distinct but $g^n = g^0$, $g^{n+1} = g^1$ and so on. Let $S = \{g\}$ and let G be the rotation group generated by S. Then G contains n distinct elements. On the other hand, the free group F generated by S contains the elements

$$\begin{array}{l} e, \quad g^1 = g, \quad g^2 = gg, \quad g^3 = ggg, \ldots, \\ \qquad g^{-1}, \quad g^{-2} = g^{-1}g^{-1}, \ldots \end{array}$$

which are all distinct. In this case, K is generated by the element g^n. Hence G is defined by one generator g and one defining relation $g^n = \mathrm{id}$. □

We have shown that the group G_3 of rotations $z \mapsto e^{i\vartheta} z$ has no canonical faithful representation in $\mathrm{SL}(2, \mathbf{C})$. More generally, let us consider a group G whose elements are transformations $g : \hat{\mathbf{C}} \to \hat{\mathbf{C}}$,

$$g(z) = \frac{az + b}{cz + d}, \quad ad - bc = 1. \tag{A.8}$$

Then

$$g(z) = w \Longleftrightarrow w = g^{-1}(z) = \frac{dw - b}{-cw + a}.$$

If $h(z) = \frac{\alpha z+\beta}{\gamma z+\delta}$, $\alpha\delta - \beta\gamma = 1$, is another element of G, then

$$g(h(z)) = \frac{(a\alpha + b\gamma)z + a\beta + b\delta}{(c\alpha + d\gamma)z + c\beta + d\delta}.$$

If g is represented by the matrix $\begin{pmatrix} a & b \\ c & d \end{pmatrix} \in \mathrm{SL}(2, \mathbf{C})$ and h by $\begin{pmatrix} \alpha & \beta \\ \gamma & \delta \end{pmatrix} \in \mathrm{SL}(2, \mathbf{C})$, then g^{-1} is represented by $\begin{pmatrix} a & b \\ c & d \end{pmatrix}^{-1}$ and $g \circ h$ by $\begin{pmatrix} a & b \\ c & d \end{pmatrix}\begin{pmatrix} \alpha & \beta \\ \gamma & \delta \end{pmatrix}$. The only problem is that both $\begin{pmatrix} a & b \\ c & d \end{pmatrix}$ and $\begin{pmatrix} -a & -b \\ -c & -d \end{pmatrix}$ represent the same mapping $g \in G$ and there is no way to define a canonical homomorphism $G \to \mathrm{SL}(2, \mathbf{C})$.

Let G be generated by $S = \{g_1, g_2, \ldots\}$. If G and the free group F generated by S are canonically isomorphic, then S generates G freely. Choose a representation

$$g_j(z) = \frac{a_j z + b_j}{c_j z + d_j}, \quad a_j d_j - b_j c_j = 1,$$

for every $g_j \in S$ and set

$$\varphi(g_j) = \begin{pmatrix} a_j & b_j \\ c_j & d_j \end{pmatrix}.$$

Since $\varphi : S \to \mathrm{SL}(2, \mathbf{C})$ has a homomorphic extension $\varphi : G \to \mathrm{SL}(2, \mathbf{C})$, G has a representation in $\mathrm{SL}(2, \mathbf{C})$. It follows that this representation is faithful. Hence we have the following theorem.

Theorem A.5.5 *If G is a free group of transformations of the form (A.8), then G has a (canonical) faithful representation in* $\mathrm{SL}(2, \mathbf{C})$. □

Let G be a group of transformations (A.8). Let G be generated by S and let R be the set of defining relations. Fix a representation

$$g(z) = \frac{az + b}{cz + d}, \quad ad - bc = 1,$$

for every $g \in S$. Let $\varphi : S \to \mathrm{SL}(2, \mathbf{C})$ be the canonical map, i.e.,

$$\varphi(g) = \begin{pmatrix} a & b \\ c & d \end{pmatrix}, \quad g \in S.$$

Let

$$r = g_1^{\varepsilon_1} \circ g_2^{\varepsilon_2} \circ \cdots \circ g_n^{\varepsilon_n} = \mathrm{id}$$

be a defining relation. Then

$$\varphi(r) = [\varphi(g_1)]^{\varepsilon_1}[\varphi(g_2)]^{\varepsilon_2} \cdots [\varphi(g_n)]^{\varepsilon_n} = \pm \begin{pmatrix} 1 & 0 \\ 0 & 1 \end{pmatrix}.$$

If $\varphi(r) = \begin{pmatrix} 1 & 0 \\ 0 & 1 \end{pmatrix}$ for all $r \in R$, then φ can be extended to an injective homomorphism $\varphi : G \to \mathrm{SL}(2, \mathbf{C})$, i.e., G has a faithful representation in $\mathrm{SL}(2, \mathbf{C})$.

A.6 Complex Möbius transformations

For geometrical reasons, a mapping $g : \hat{\mathbf{C}} \to \hat{\mathbf{C}}$ which can be represented in the form

$$g(z) = \frac{az + b}{cz + d}, \quad \begin{pmatrix} a & b \\ c & d \end{pmatrix} \in \mathrm{GL}(2, \mathbf{C}), \tag{A.9}$$

is called a (*complex*) *Möbius transformation*. Note that $g(-d/c) = \infty$ and $g(\infty) = a/c$. It follows that g is a homeomorphism. The group of Möbius transformations is denoted by $M = M(\mathbf{C})$.

Theorem A.6.1 *If z_1, z_2, $z_3 \in \hat{\mathbf{C}}$ are distinct points and w_1, w_2, $w_3 \in \hat{\mathbf{C}}$ are distinct points, then there exists a unique $g \in M$ for which $g(z_j) = w_j$, $j = 1, 2, 3$.* □

Define a mapping

$$\Phi : \mathrm{GL}(2, \mathbf{C}) \to M$$

as follows: If $A = \begin{pmatrix} a & b \\ c & d \end{pmatrix} \in \mathrm{GL}(2, \mathbf{C})$, then $\Phi(A) = g_A$ where

$$g_A(z) = \frac{az + b}{cz + d}.$$

Since $g_A \circ g_B = g_{AB}$, Φ is a homomorphism and

$$A \in \ker \Phi \iff \frac{az + b}{cz + d} = z \quad \text{for all } z \in \hat{\mathbf{C}}.$$

By choosing $z = 0, \infty, 1$ we obtain $b = 0$, $c = 0$, $a/d = 1$. Hence

$$A \in \ker \Phi \iff A = \begin{pmatrix} a & 0 \\ 0 & a \end{pmatrix}, \quad a \neq 0.$$

If $K = \ker \Phi$, then

$$M(\mathbf{C}) \simeq \mathrm{GL}(2, \mathbf{C})/K.$$

Denote $\Phi_0 = \Phi|\mathrm{SL}(2,\mathbf{C})$ and

$$K_0 = \ker \Phi_0 = K \cap \mathrm{SL}(2,\mathbf{C}) = \{-I, I\}.$$

Then every $g \in M$ is the projection of exactly two matrices A, $-A \in \mathrm{SL}(2,\mathbf{C})$. Moreover,

$$M \simeq \mathrm{SL}(2,\mathbf{C})/K_0 = \mathrm{PSL}(2,\mathbf{C}).$$

If $A = \begin{pmatrix} a & b \\ c & d \end{pmatrix} \in \mathrm{GL}(2,\mathbf{C})$ and $\lambda \neq 0$, then

$$\frac{\mathrm{tr}^2 A}{\det A} = \frac{\mathrm{tr}^2(\lambda A)}{\det(\lambda A)},$$

i.e., the function $A \mapsto (\mathrm{tr}^2 A)/\det A$ is invariant under the mappings $A \mapsto \lambda A$. Similarly, also the function $\|A\|^2/|\det A|$ is invariant. Hence we have in M well-defined functions

$$\mathrm{tr}^2 g = \frac{\mathrm{tr}^2 A}{\det A} \quad \text{and} \quad \|g\| = \frac{\|A\|}{|\det A|^{1/2}}, \quad \Phi(A) = g.$$

Denote $|\mathrm{tr}\, g| = |\mathrm{tr}^2 g|^{1/2} \geq 0$.

Let $G \subset M$ be a group. The *stabilator*

$$G_z = \{\, g \in G \mid g(z) = z \,\}$$

of $z \in \hat{\mathbf{C}}$ is a subgroup of G. The *G–orbit*

$$G(z) = \{\, g(z) \mid g \in G \,\}$$

of $z \in \hat{\mathbf{C}}$ is a subset of $\hat{\mathbf{C}}$. Two points z_1, $z_2 \in \hat{\mathbf{C}}$ are *equivalent* under G if $g(z_1) = z_2$ for some $g \in G$. Two subgroups G_1 and G_2 of M are *conjugate* if $hG_1h^{-1} = G_2$ for some $h \in M$.

Theorem A.6.2 *The following conditions are equivalent:*

- *z_1 and z_2 are equivalent under G,*
- *$G(z_1) = G(z_2)$,*
- *There exists $w \in \hat{\mathbf{C}}$ such that z_1, $z_2 \in G(w)$,*
- *G_{z_1} and G_{z_2} are conjugate.* □

Let G and G' be isomorphic subgroups of M, and let $j : G \to G'$ be an isomorphism. If there exists a homeomorphism $\varphi : \hat{\mathbf{C}} \to \hat{\mathbf{C}}$ such that

$$j(g) = \varphi \circ g \circ \varphi^{-1}$$

for all $g \in G$, then j is *geometric*. Note that also $j^{-1} : G' \to G$ is geometric. If $\varphi \in M$, then j is a *conjugation* and the transformations g and $j(g) = \varphi \circ g \circ \varphi^{-1}$ are conjugate. If two groups are geometrically isomorphic, they share many geometrical properties. For instance, if z is a fixed point of $g \in G$, then $\varphi(z)$ is a fixed point of $j(g)$.

Example. Let G be generated by $g : z \mapsto z + 1$ and G' by $g' : z \mapsto 3z$. Since G and G' are cyclic free Abelian groups, the mapping $j : g \mapsto g'$ has an isomorphic extension $j : G \to G'$. This isomorphism cannot be geometric since g has only one fixed point but g' has two fixed points in $\hat{\mathbf{C}}$. □

Let F_g be the set of the fixed points of $g \in M$. Then evidently $F_g = F_{g^{-1}}$.

Theorem A.6.3 *Let g, $h \in M$. If g and h commute, i.e., if $g \circ h \circ g^{-1} \circ h^{-1} = \mathrm{id}$, then $g(F_h) = F_h$ and $h(F_g) = F_g$.*

Proof. If $z \in F_h$, then $h(g(z)) = g(h(z)) = g(z)$. Thus $g(F_h) \subset F_h$. Similarly, since $h^{-1}(g^{-1}(z)) = (g \circ h)^{-1}(z) = (h \circ g)^{-1}(z) = g{-}1(h^{-1}(z)) = g^{-1}(z)$, we have $g^{-1}(F_h) \subset F_h$. Hence $F_h = g(F_h)$. □

Examples. 1. Let $g(z) = 1/z$ and $h(z) = kz$, $k > 1$. Then

$$F_h = \{0, \infty\} \quad \text{and} \quad F_g = \{-1, 1\}.$$

We have $g(F_h) = F_h$ but $h(F_g) \neq F_g$. On the other hand, $g \circ h = h^{-1} \circ g \neq h \circ g$. In fact, $h \circ g = g \circ h$ if and only if $g(F_h) = F_h$ and $h(F_g) = F_g$ (cf. Theorem A.7.3).

2. Let $g(z) = 1/z$ and $h(z) = -z$. Then

$$F_h = \{0, \infty\} \quad \text{and} \quad F_g = \{-1, 1\}.$$

We have $g(F_h) = F_h$ and $h(F_g) = F_g$. On the other hand, $g \circ h = h \circ g$. Note that $F_h \cap F_g = \emptyset$. □

Let $g \in M$,

$$g(z) = \frac{az + b}{cz + d}, \quad ad - bc = 1.$$

Then $z \in F_g$ if and only if $cz^2 - (a - d)z - b = 0$. We have three alternatives:

- $F_g = \hat{\mathbf{C}}$ and $g = \mathrm{id}$, i.e., $a - d = c = b = 0$,

- F_g contains one point only, i.e., $(a+d)^2 - 4 = 0$,
- F_g contains two points, i.e., $(a+d)^2 - 4 \neq 0$.

Example. Let $g(z) = kz$, $k > 1$, and $h(z) = -z$. Then $F_g = F_h = \{0, \infty\}$. However, the mapping properties of g and h are quite different. E.g. $h^2 = \mathrm{id}$ but $g^2 \neq \mathrm{id}$. Hence we have to improve the above classification of Möbius transformations. □

Define $m_k \in M$ by setting

$$\begin{aligned} m_k(z) &= kz \quad \text{for } k \in \mathbf{C},\ k \neq 0, 1, \\ m_1(z) &= z + 1. \end{aligned}$$

Theorem A.6.4 $\mathrm{tr}^2 m_k = k + k^{-1} + 2$.

Proof. If generally $g = \Phi(A)$, $A \in \mathrm{GL}(2, \mathbf{C})$, then

$$\mathrm{tr}^2 g = \frac{\mathrm{tr}^2 A}{\det A}.$$

If $k \neq 1$, then $A = \left(\begin{smallmatrix} k & 0 \\ 0 & 1 \end{smallmatrix}\right)$ and

$$\mathrm{tr}^2 m_k = \frac{(k+1)^2}{k} = k + k^{-1} + 2.$$

If $k = 1$, then $A = \left(\begin{smallmatrix} 1 & 1 \\ 0 & 1 \end{smallmatrix}\right)$ and

$$\mathrm{tr}^2 m_1 = \frac{(1+1)^2}{1} = 4. \ \square$$

Theorem A.6.5 *If A, $B \in \mathrm{SL}(2, \mathbf{C})$, then* $\mathrm{tr}\, A = \mathrm{tr}(BAB^{-1})$.

Proof. Let $A = \left(\begin{smallmatrix} a & b \\ c & d \end{smallmatrix}\right)$ and $B = \left(\begin{smallmatrix} \alpha & \beta \\ \gamma & \delta \end{smallmatrix}\right)$. Then

$$\begin{aligned} BAB^{-1} &= \begin{pmatrix} \alpha & \beta \\ \gamma & \delta \end{pmatrix} \begin{pmatrix} a & b \\ c & d \end{pmatrix} \begin{pmatrix} \delta & -\beta \\ -\gamma & \alpha \end{pmatrix} \\ &= \begin{pmatrix} \alpha a + \beta c & \alpha b + \beta d \\ \gamma a + \delta c & \gamma b + \delta d \end{pmatrix} \begin{pmatrix} \delta & -\beta \\ -\gamma & \alpha \end{pmatrix} \\ &= \begin{pmatrix} \alpha\delta a + \beta\delta c - \alpha\gamma b - \beta\gamma d & \bullet \\ \bullet & -\beta\gamma a - \beta\delta c + \alpha\gamma b + \alpha\delta d \end{pmatrix} \end{aligned}$$

and hence

$$\operatorname{tr}(BAB^{-1}) = (\alpha\delta - \beta\gamma)a + (\alpha\delta - \beta\gamma)d = \operatorname{tr} A. \square$$

Let g, $h \in M$ and choose A, $B \in \mathrm{SL}(2, \mathbf{C})$ such that $\Phi(A) = g$ and $\Phi(B) = h$. Then $\Phi(BAB^{-1}) = h \circ g \circ h^{-1}$. Hence we have the following corollary for Theorem A.6.5.

Corollary A.6.6 $\operatorname{tr}^2 g = \operatorname{tr}^2(h \circ g \circ h^{-1})$ *for all* g, $h \in M$. $\square$

Theorem A.6.7 *Let g be a Möbius transformation with two fixed points x and y. Choose $h \in M$ such that $h(x) = \infty$ and $h(y) = 0$. Then $h \circ g \circ h^{-1} = m_k$ where the number $k \neq 1$ does not depend on the choice of h.*

Proof. We have $F_{h \circ g \circ h^{-1}} = \{0, \infty\} = F_{m_k}$ for all $k \neq 1$. Let $k = h(g(h^{-1}(1)))$. Then $k \neq 1$ and $m_k(1) = k$. Hence $m_k^{-1} \circ (h \circ g \circ h^{-1})$ has 1, 0 and ∞ as fixed points. By Theorem A.6.1, $m_k = h \circ g \circ h^{-1}$.

If $m_{k_0} = h_0 \circ g \circ h_0^{-1}$, $h_0(x) = \infty$ and $h_0(y) = 0$, then $m_k = (h \circ h_0^{-1}) \circ m_{k_0} \circ (h \circ h_0^{-1})^{-1}$. Since $F_{h \circ h_0^{-1}} = \{0, \infty\}$, we have $h \circ h_0^{-1} = m_{k'}$, $k' \neq 1$. Then $m_{k_0} = m_{k'} \circ m_k \circ (m_{k'})^{-1} = m_k$. $\square$

Theorem A.6.8 *Let g be a Möbius transformation having one fixed point x only. Let $y \neq x$ and choose $h \in M$ such that $h(x) = \infty$, $h(y) = 0$ and $h(g(y)) = 1$. Then $h \circ g \circ h^{-1} = m_1$.*

Proof. Since g and $h \circ g \circ h^{-1}$ are conjugate, also $h \circ g \circ h^{-1}$ has one fixed point only, i.e., $F_{h \circ g \circ h^{-1}} = \{\infty\}$. Hence $h \circ g \circ h^{-1} : z \mapsto z + b$. Since

$$1 = h(g(y)) = h(g(h^{-1}(0))) = b,$$

we have $h \circ g \circ h^{-1} = m_1$. $\square$

By Theorems A.6.7 and A.6.8, every Möbius transformation is conjugate to a *standard transformation* m_k. Moreover, all transformations with exactly one fixed point are conjugate to $m_1 : z \mapsto z + 1$. Suppose that a Möbius transformation g with two fixed points is conjugate to standard transformations m_p and m_q. Then also m_p and m_q are conjugate, and the next theorem shows that either $m_p = m_q$ or $m_p = m_q^{-1}$.

Theorem A.6.9 *Standard transformations m_p and m_q are conjugate if and only if $p = q$ or $p = q^{-1}$.*

Proof. Suppose that m_p and m_q are conjugate. Then by Corollary A.6.6, $\mathrm{tr}^2 m_k = \mathrm{tr}^2 m_q$, and we have by Theorem A.6.4

$$p + p^{-1} = q + q^{-1},$$

i.e., either $p = q$ or $p = q^{-1}$.

Suppose conversely that either $p = q$ or $p = q^{-1}$. In the first case, $m_p = m_q$. If $p = q^{-1} \neq q$, let $h(z) = z^{-1}$. Then

$$h(m_p(h^{-1}(z))) = (pz^{-1})^{-1} = p^{-1}z = m_q(z)$$

and m_p and m_q are conjugate. □

Theorem A.6.10 *Let $g_1, g_2 \in M \setminus \{\mathrm{id}\}$. Then g_1 and g_2 are conjugate if and only if $\mathrm{tr}^2 g_1 = \mathrm{tr}^2 g_2$.*

Proof. If g_1 and g_2 are conjugate, then $\mathrm{tr}^2 g_1 = \mathrm{tr}^2 g_2$ by Corollary A.6.6.

Suppose that $\mathrm{tr}^2 g_1 = \mathrm{tr}^2 g_2$. Let g_1 be conjugate to standard transformation m_p and g_2 to standard transformation m_q. Then, by Corollary A.6.6,

$$\mathrm{tr}^2 m_p = \mathrm{tr}^2 g_1 = \mathrm{tr}^2 g_2 = \mathrm{tr}^2 m_q.$$

Hence, by Theorem A.6.4, $p + p^{-1} = q + q^{-1}$, i.e., either $p = q$ or $p = q^{-1}$. By Theorem A.6.9, m_p and m_q are conjugate and hence also g_1 and g_2 are conjugate. □

Theorem A.6.11 *Suppose that $g \in M$ has two finite fixed points x and y. Then*

$$g(z) = \frac{(kx - y)z - xy(k-1)}{(k-1)z + x - ky} \tag{A.10}$$

and g and standard transformation m_k are conjugate.

Proof. Let $h(z) = \frac{z-y}{z-x}$. Then $h \in M$, $h(x) = \infty$ and $h(y) = 0$. Denote $m_k = h \circ g \circ h^{-1}$. Since $h \circ g = m_k \circ h$, we have

$$\frac{g(z) - y}{g(z) - x} = k\frac{z - y}{z - x}$$

from which the assertion follows. □

If $x = \infty$, then we obtain from (A.10) by letting $x \to \infty$

$$g(z) = kz - y(k-1).$$

Similarly

$$g(z) = \frac{z}{k} + x\left(1 - \frac{1}{k}\right)$$

if $y = \infty$.

A standard transformation m_p, $p > 1$, maps the circle $|z| = r$ onto the circle $|z| = pr$. On the other hand, if $|q| = 1$ and $q \neq 1$, then m_q maps every circle $|z| = r$ onto itself. This basic distinction gives rise to the following classification of the Möbius transformations.

Definition A.6.1 *Let $g \in M \setminus \{\mathrm{id}\}$. Then*

- *g is* parabolic *if g has only one fixed point in $\hat{\mathbf{C}}$, i.e., g is conjugate to m_1,*
- *g is* loxodromic *if g has two fixed points in $\hat{\mathbf{C}}$ and g is conjugate to m_k, $|k| \neq 1$,*
- *g is* elliptic *if g has two fixed points in $\hat{\mathbf{C}}$ and g is conjugate to m_k, $|k| = 1$.*

By Theorem A.6.9, the *type* of g given in Definition A.6.1 does not depend on the choice of m_k. Moreover, conjugate transformations are always of the same type.

If $|k| > 1$ and $z \neq 0, \infty$, then

$$m_k^n(z) \to \begin{cases} 0 & \text{as } n \to -\infty, \\ \infty & \text{as } n \to \infty. \end{cases}$$

Definition A.6.2 *Let g be loxodromic and $z \notin F_g$. Then*

$$a(g) = \lim_{n\to\infty} g^n(z)$$

is the attractive *fixed point of g and*

$$r(g) = \lim_{n\to\infty} g^{-n}(z)$$

is the repulsive *fixed point of g.*

A domain $D \subset \hat{\mathbf{C}}$ is a *disk* if it is bounded by a circle. A half–plane is a disk whose boundary circle passes through ∞. It can be shown by elementary calculation that disks are mapped onto disks by Möbius transformations. Let $g \in M$. We are interested in finding a disk D such that $g(D) = D$. Suppose that $g = m_k$, $k \neq |k| > 1$. Then $g(D) \neq D$ for any disk $D \subset \hat{\mathbf{C}}$. On the other hand, if $g = m_k$ with $k > 1$, then $g(D) = D$ if and only if D is a half–plane bounded by a straight line through the origin.

Definition A.6.3 *Let $g \in M$ be loxodromic. If $g(D) = D$ for some disk $D \subset \hat{\mathbf{C}}$, then g is* hyperbolic. *Otherwise g is* strictly loxodromic.

By definition, m_k is parabolic if and only if $k = 1$, i.e., $\operatorname{tr}^2 m_k = 4$. Similarly, m_k is elliptic if and only if $k = e^{i\vartheta}$, $\vartheta \in \mathbf{R}$. Then $\operatorname{tr}^2 m_k = e^{i\vartheta} + e^{-i\vartheta} = 2 + 2\cos\vartheta$. Hence m_k is elliptic if and only if $0 \le \operatorname{tr}^2 m_k < 4$. If m_k is loxodromic, then m_k is hyperbolic if and only if $k > 0$, $k \ne 1$, i.e., $\operatorname{tr}^2 m_k > 4$. The following theorem can now be proved by elementary geometrical reasoning.

Theorem A.6.12 *Let $g \in M \setminus \{\mathrm{id}\}$. Then*

- *g is parabolic if and only if $\operatorname{tr}^2 g = 4$,*
- *g is elliptic if and only if $0 \le \operatorname{tr}^2 g < 4$,*
- *g is hyperbolic if and only if $\operatorname{tr}^2 g > 4$,*
- *g is strictly loxodromic if and only if $\operatorname{tr}^2 g \ne |\operatorname{tr}^2 g|$.* □

The *cross–ratio* (z_1, z_2, z_3, z_4) of distinct points $z_j \in \hat{\mathbf{C}}$, $i = 1, 2, 3, 4$, is defined as the image of z_1 under the Möbius transformation for which $z_2 \mapsto 1$, $z_3 \mapsto 0$ and $z_4 \mapsto \infty$. If all points are finite, then

$$(z_1, z_2, z_3, z_4) = \frac{z_1 - z_3}{z_1 - z_4} \cdot \frac{z_2 - z_4}{z_2 - z_3}.$$

If $g \in M$, then by Theorem A.6.1

$$(z_1, z_2, z_3, z_4) = (g(z_1), g(z_2), g(z_3), g(z_4)).$$

Let $g \in M$ have two fixed points. Then it follows similarly as in the proof of Theorem A.6.11 that the cross–ratio

$$(g(z), z, x, y) = k$$

does not depend on the choice of the point $z \in \hat{\mathbf{C}} \setminus \{x, y\}$. Moreover, g and m_k are conjugate.

Suppose that g is loxodromic. Choose $x = a(g)$ and $y = r(g)$. Denote

$$k(g) = (g(z), z, x, y).$$

We have $|k(g)| > 1$ by Definition A.6.2. Let g be elliptic. By Theorem A.6.9, we may choose x and y such that

$$k(g) = (g(z), z, x, y) = e^{i\vartheta}, \quad 0 < \vartheta \le \pi.$$

The number $k(g)$ is called *the multiplier* of a loxodromic or elliptic transformation g. For a parabolic g, we set $k(g) = 1$.

Theorem A.6.13 *Let $g \in M \setminus \{\mathrm{id}\}$. Then*

- *g is hyperbolic if and only if $k(g) > 1$,*
- *g is strictly loxodromic if and only if $k(g) \neq |k(g)| > 1$.*

Proof. Theorems A.6.4 and A.6.12. □

A.7 Abelian groups of Möbius transformations

Consider Möbius transformations g and h and matrices A, $B \in \mathrm{SL}(2,\mathbf{C})$ representing them, i.e., $g = \Phi(A)$ and $h = \Phi(B)$. Since also $-A$ and $-B$ represent g and h, respectively, the trace of a Möbius transformation is determined only up to the sign. Therefore, the square of the trace is a well–defined function in M. However, we have the following result.

Theorem A.7.1 *The trace*

$$\mathrm{tr}(g \circ h \circ g^{-1} \circ h^{-1}) = \mathrm{tr}(ABA^{-1}B^{-1})$$

is uniquely determined.

Proof. The matrix product $ABA^{-1}B^{-1}$ does not change if A is replaced by $-A$ or B is replaced by $-B$. □

By the proof of Theorem A.7.1, we may speak about *the* matrix of the commutator $g \circ h \circ g^{-1} \circ h^{-1}$ in $\mathrm{SL}(2,\mathbf{C})$.

Theorem A.7.2 (i) *Möbius transformations g and h have a common fixed point if and only if* $\mathrm{tr}(g \circ h \circ g^{-1} \circ h^{-1}) = 2$.
(ii) *If $g \neq \mathrm{id}$ and $h \neq \mathrm{id}$ share a fixed point, then either*

(a) *$g \circ h = h \circ g$ and $F_g = F_h$, or*

(b) *$g \circ h \circ g^{-1} \circ h^{-1}$ is parabolic and $F_g \neq F_h$.*

Proof. In (i), we may suppose that $g(\infty) = \infty$. Let

$$g = \begin{pmatrix} a & b \\ 0 & d \end{pmatrix}, \quad ad = 1, \quad \text{and}$$

$$h = \begin{pmatrix} \alpha & \beta \\ \gamma & \delta \end{pmatrix}, \quad \alpha\delta - \beta\gamma = 1.$$

Then, by a lengthy computation,

$$t = \mathrm{tr}(g \circ h \circ g^{-1} \circ h^{-1}) = 2 + b^2\gamma^2 + b(a-d)\gamma(\alpha - \delta) - (a-d)^2\gamma\beta$$

Suppose that g and h share a fixed point. We may suppose that $\infty \in F_g \cap F_h$. Then $\gamma = 0$ and $t = 2$.

Conversely, suppose that $t = 2$. If g is a parabolic, then $g(z) = z + b$ and $a - d = 0$. Hence $t = 2 + b^2\gamma^2 = 2$ and it follows that $\gamma = 0$ and $\infty \in F_g \cap F_h$. If g is not parabolic, then we may suppose that $F_g = \{0, \infty\}$, i.e., $b = 0$ and $ad = 1$, $a \neq d$. Then

$$t = 2 - (a-d)^2\gamma\beta = 2$$

and we have $\gamma\beta = 0$. Hence $F_g \cap F_h \neq \emptyset$.

In (ii), we may suppose $\infty \in F_g \cap F_h$. Then by (i), either $g \circ h \circ g^{-1} \circ h^{-1} = \mathrm{id}$ or $g \circ h \circ g^{-1} \circ h^{-1}$ is parabolic. Moreover, $g \circ h \circ g^{-1} \circ h^{-1}$ fixes ∞. Hence $g \circ h \circ g^{-1} \circ h^{-1} = \mathrm{id}$ if and only if $g(h(g^{-1}(h^{-1}(0)))) = 0$. Since

$$g = \begin{pmatrix} a & b \\ 0 & d \end{pmatrix} \quad \text{and} \quad h = \begin{pmatrix} \alpha & \beta \\ 0 & \delta \end{pmatrix},$$

we have $g(h(g^{-1}(h^{-1}(0)))) = (-d\beta - b\alpha + a\beta + b\delta)/d\delta$. Hence $g \circ h \circ g^{-1} \circ h^{-1} = \mathrm{id}$ if and only if $\beta(a - d) = b(\alpha - \delta)$.

Suppose that $a - d = 0$. Since $ad = 1$, we may suppose that $a = d = 1$. From $g \neq \mathrm{id}$ it then follows that $b \neq 0$. Hence $\beta(a-d) = b(\alpha - \delta)$ if and only if $\alpha = \delta = \pm 1$ and $\beta \neq 0$, i.e., $F_g = F_h = \{\infty\}$.

Suppose that $a - d \neq 0$. Then g has two fixed points. We may suppose that $F_g = \{0, \infty\}$. Since $b = 0$, we have $\beta(a - d) = b(\alpha - \delta)$ if and only if $\beta = 0$, i.e., $F_g = F_h = \{0, \infty\}$. □

If $g, h \in M \setminus \{\mathrm{id}\}$ commute, then $g(F_h) = F_h$ and $h(F_g) = F_g$ by Theorem A.6.3. Also the converse holds:

Theorem A.7.3 *Let $g, h \in M \setminus \{\mathrm{id}\}$. Then $g \circ h = h \circ g$ if and only if $g(F_h) = F_h$ and $h(F_g) = F_g$.*

Proof. Suppose that $g(F_h) = F_h$ and $h(F_g) = F_g$. It suffices to show that $g \circ h = h \circ g$.

We have either $F_g = F_h$ or $F_g \cap F_h = \emptyset$. Indeed, if we had

$$y \in F_g \cap F_h \quad \text{and} \quad x \in F_g \setminus F_h \quad (\text{or } x \in F_h \setminus F_g),$$

then the condition $h(F_g) = F_g$ would imply that F_g would contain three distinct points x, y and $h(x)$ which is impossible.

Suppose that g is parabolic. We may suppose that $F_g = \{\infty\}$ and $g(z) = z+1$. Since $h(F_g) = F_g$, we have $\infty \in F_g \cap F_h$ and hence $F_g = F_h = \{\infty\}$. Since $h(z) = z + \beta$ for some $\beta \neq 0$, we have $g \circ h = h \circ g$.

Suppose that g has two fixed points. We may suppose that $F_g = \{0, \infty\}$. Then $g(z) = kz$, $k \neq 1$. If $F_h = F_g$, then $h(z) = k'z$ for some $k' \neq 1$ and we have $g \circ h = h \circ g$.

It remains the case $F_g \cap F_h = \emptyset$. Let $h(z) = \frac{az+b}{cz+d}$, $ad - bc = 1$. Since $h(F_g) = F_g$, we have $h(0) = \infty$ and $h(\infty) = 0$, i.e., $a = d = 0$. Hence $h(z) = \alpha/z$, $\alpha = b/c \neq 0$. Then $F_h = \{\sqrt{\alpha}, -\sqrt{\alpha}\}$. Since $g(F_h) = F_h$, $g(z) = kz$, we have $k = -1$. Hence

$$g(z) = -z \quad \text{and} \quad h(z) = \frac{\alpha}{z}$$

and we have $g \circ h = h \circ g$. □

If $g \circ h = h \circ g$ and $F_g \cap F_h = \emptyset$, then by the above proof,

$$g^2 = h^2 = (g \circ h)^2 = (h \circ g)^2 = \mathrm{id},$$

i.e., all these mappings are elliptic and of order two. Note that g is of order two if and only if $\mathrm{tr}^2 g = 0$.

The transformations id, $g(z) = -z$, $h(z) = \alpha/z$ and $g(h(z)) = -\alpha/z$ constitute an Abelian group $G \subset M$. Conjugating G by the transformation $z \mapsto z/\sqrt{\alpha}$, we obtain the *quadratic group* which contains the transformations id, $z \mapsto -z$, $z \mapsto 1/z$ and $z \mapsto -1/z$.

Theorem A.7.4 *Suppose that $G \subset M$ is an Abelian group. Then either $F_g = F_h$ for all g, $h \in G \setminus \{\mathrm{id}\}$ or G is conjugate to the quadratic group.*

Proof. Suppose that there exists g, $h \in G \setminus \{\mathrm{id}\}$ such that $F_g \neq F_h$. By the proof of Theorem A.6.4, $F_g \cap F_h = \emptyset$ and g and h have both two fixed points. Moreover, if we normalize such that $F_g = \{0, \infty\}$, then

$$g(z) = -z \quad \text{and} \quad h(z) = \frac{\alpha}{z}, \quad \alpha \neq 0.$$

Suppose that $h' \in G \setminus \{g, \mathrm{id}\}$. Then, by the above reasoning, $h'(z) = \alpha'/z$, $\alpha' \neq 0$. Since $h' \circ h = h \circ h'$, we have $(\alpha')^2 = \alpha^2$, i.e., $\alpha' = \pm\alpha$. Hence either $h' = h$ or $h' = g \circ h$. □

If $G \subset M$ is a group such that $F_g = F_h$ for all g, $h \in G \setminus \{\mathrm{id}\}$, then G is Abelian by Theorem A.7.3.

Suppose that $G \subset M$ is a group consisting of the identity and three elliptic transformations g_1, g_2 and g_3 of order two. If $F_{g_i} = F_{g_j}$, then

$g_i = g_j$. Hence by Theorem A.7.2, $F_{g_i} \cap F_{g_j} = \emptyset$ for all $i \neq j$. Moreover, $g_i \circ g_j = g_k$ and $g_j \circ g_i = g_k$, $\{i,j,k\} = \{1,2,3\}$. Hence G is conjugate to the quadratic group.

A group $G \subset M$ is *purely elliptic* if it contains besides the identity elliptic elements only. For example, the quadratic group is purely elliptic. We show next that all purely elliptic groups can be interpreted as groups of spherical rotations.

Consider the $x_1x_2x_3$–space and choose the x_1x_2–plane as the complex plane $\mathbf{C}$. Let S be the sphere of radius $1/2$ centered at $(0,0,1/2)$. Stereographic projection p with $(0,0,1)$ as the projection center maps S bijectively onto $\hat{\mathbf{C}}$ if $(0,0,1)$ and ∞ are let to correspond to each other. Hence S and $\hat{\mathbf{C}}$ can be identified and $S = \hat{\mathbf{C}}$ is called the *Riemann sphere*.

The stereographic projection p maps the Riemann sphere conformally onto extended complex plane.

Let K be a rotation of the Riemann sphere. Then

$$g = p \circ K \circ p^{-1}$$

maps the extended complex plane bijectively and conformally onto itself. By a well–known result in complex analysis, g is a Möbius transformation. By geometry, g has two fixed points. Since a spherical rotation has no attracting fixed points, g is elliptic.

Let x and y be two points in $\hat{\mathbf{C}}$. Denote by s_x and s_y the points of the Riemann sphere S for which $p(s_x) = x$ and $p(s_y) = y$. It can be shown by geometrical reasoning that s_x and s_y are antipodal points of S if and only if $y = -1/\overline{x}$. These slightly heuristical considerations justify the following definition:

Definition A.7.1 *A transformation $g \in M$ is a* spherical rotation *if either $g = \mathrm{id}$ or g is elliptic and the fixed points x and y of g satisfy $y = -1/\overline{x}$.*

The standard transformations m_k, $k = e^{i\varphi}$, are spherical rotations; these transformations are represented by matrices of the form

$$\begin{pmatrix} \varepsilon & 0 \\ 0 & \overline{\varepsilon} \end{pmatrix}, \quad |\varepsilon| = 1. \qquad \text{(A.11)}$$

The quadratic group contains besides the identity three distinct spherical rotations whose rotation axes are perpendicular to each other.

Theorem A.7.5 *A transformation $g \in M$ is a spherical rotation if and only if g is of the form*

$$g(z) = \frac{az+b}{-\overline{b}z+\overline{a}}, \quad |a|^2 + |b|^2 = 1.$$

Proof. Let $g \in M \setminus \{\mathrm{id}\}$ be a spherical rotation. Then g is elliptic. If one of the fixed points x and y is ∞, then the other is 0, and the assertion follows by (A.11).

Suppose that x and y are finite. Then by (A.10)

$$g = \begin{pmatrix} a & b \\ c & d \end{pmatrix} = \begin{pmatrix} \frac{kx-y}{(x-y)\sqrt{k}} & -\frac{xy(k-1)}{(x-y)\sqrt{k}} \\ \frac{k-1}{(x-y)\sqrt{k}} & \frac{x-ky}{(x-y)\sqrt{k}} \end{pmatrix}, \quad ad - bc = 1.$$

Since $y = -1/\overline{x}$, we obtain

$$a = \frac{kx\overline{x} + 1}{(x\overline{x} + 1)\sqrt{k}}, \quad d = \frac{x\overline{x} + k}{(x\overline{x} + 1)\sqrt{k}}.$$

On the other hand, we have $|k|^2 = 1$. Hence $\overline{k} = 1/k$ and

$$\overline{a} = \frac{\frac{1}{k}x\overline{x} + 1}{(x\overline{x} + 1)\sqrt{\frac{1}{k}}} = \frac{x\overline{x} + k}{(x\overline{x} + 1)\sqrt{k}} = d.$$

Similarly,

$$b = \frac{x(k-1)}{(x\overline{x} + 1)\sqrt{k}}, \quad c = \frac{\overline{x}(k-1)}{(x\overline{x} + 1)\sqrt{k}},$$

and

$$-\overline{b} = -\frac{\overline{x}(\frac{1}{k} - 1)}{(x\overline{x} + 1)\sqrt{\frac{1}{k}}} = -\frac{\overline{x}(1-k)}{(x\overline{x} + 1)\sqrt{k}} = c.$$

Conversely, suppose that

$$g = \begin{pmatrix} a & b \\ -\overline{b} & \overline{a} \end{pmatrix}, \quad |a|^2 + |b|^2 = 1.$$

Then $\mathrm{tr}^2 g = (a + \overline{a})^2 = 4(\mathrm{Re}\, a)^2 \leq 4$. If $\mathrm{Re}\, a = \pm 1$, then $a = \pm 1$ and $b = 0$ and $g = \mathrm{id}$. Otherwise $\mathrm{tr}^2 g < 4$ and g is elliptic.

Suppose that $g \neq \mathrm{id}$. If $b = 0$, g is a spherical rotation by (A.11). Hence we may suppose that $b \neq 0$. The fixed points x and y of g are obtained by solving the equation $\overline{b}z^2 - (\overline{a} - a)z + b = 0$. Hence

$$x = \frac{\overline{a} - a + \sqrt{(a + \overline{a})^2 - 4}}{2\overline{b}}$$
$$y = \frac{\overline{a} - a - \sqrt{(a + \overline{a})^2 - 4}}{2\overline{b}}.$$

The numerator of y is purely imaginary. Hence

$$\overline{y} = \frac{-(\overline{a} - a) + \sqrt{(a + \overline{a})^2 - 4}}{2b}$$

and we have

$$x\overline{y} = \frac{(a+\overline{a})^2 - 4 - (\overline{a}-a)^2}{4b\overline{b}} = \frac{|a|^2-1}{|b|^2} = \frac{-|b|^2}{|b|^2} = -1. \square$$

Geometrically it is evident that spherical rotations constitute a group. This fact is not so evident if we consider Definition A.7.1 only but it follows immediately from Theorem A.7.5. Denote by $\mathcal{K}$ the group of spherical rotations.

Theorem A.7.6 *If G is a purely elliptic subgroup of M, then $hGh^{-1} \subset \mathcal{K}$ for some $h \in M$.*

Proof. We use matrix notation for Möbius transformations. Hence matrices will denote elements of $\mathrm{PSL}(2,\mathbf{C})$, i.e., they are determined up to the factor $\pm I$.

After a suitable conjugation G contains a spherical rotation

$$U = \begin{pmatrix} \varepsilon & 0 \\ 0 & \overline{\varepsilon} \end{pmatrix}, \quad |\varepsilon| = 1,\ \varepsilon \neq \pm 1,$$

whose fixed points are 0 and ∞.

Let $V \in G \setminus \{\mathrm{id}\}$,

$$V = \begin{pmatrix} a & b \\ c & d \end{pmatrix}, \quad ad - bc = 1.$$

If $F_V = \{0, \infty\}$, then $V \in \mathcal{K}$. If $F_U \cap F_V$ contains exactly one point, then $UVU^{-1}V^{-1}$ is parabolic by Theorem A.7.2. Hence we may suppose that $V(0) \neq 0$ and $V(\infty) \neq \infty$. Then $UV \neq \mathrm{id}$. Since V and UV are elliptic, the traces $\mathrm{tr}\, V = a + d$ and $\mathrm{tr}\,(UV) = \varepsilon a + \overline{\varepsilon} d$ are real. Then $d = \overline{a} + r$ where $r = a + d - (a + \overline{a})$ is real. From $\varepsilon a + \overline{\varepsilon} d = \varepsilon a + \overline{\varepsilon}\,\overline{a} + \overline{\varepsilon} r$ it follows also $\overline{\varepsilon} r$ is real. Since $\overline{\varepsilon} \neq \pm 1$, we have $r = 0$ and $d = \overline{a}$. Hence

$$V = \begin{pmatrix} a & b \\ c & \overline{a} \end{pmatrix}, \quad a\overline{a} - bc = 1. \tag{A.12}$$

The commutator $C = UVU^{-1}V^{-1}$ is either the identity or elliptic. Hence (cf. the proof of Theorem A.7.2)

$$-2 \leq \mathrm{tr}\, C = 2 - bc(\varepsilon - \overline{\varepsilon})^2 \leq 2.$$

Since $(\varepsilon - \overline{\varepsilon})^2 < 0$, we have $bc \leq 0$. On the other hand, it follows from the conditions $V(0) \neq 0$ and $V(\infty) \neq \infty$ that $bc \neq 0$. Thus $bc < 0$.

By Theorem A.7.5, $V \in \mathcal{K}$ if and only if $c = -\overline{b}$. We may conjugate by any transformation $A \in M$ of the form

$$A = \begin{pmatrix} \mu & 0 \\ 0 & \mu^{-1} \end{pmatrix}, \quad \mu > 0,$$

since $AUA^{-1} = U$ by Theorem A.7.3. Now

$$AVA^{-1} = \begin{pmatrix} a & b\mu^2 \\ c\mu^{-2} & \overline{a} \end{pmatrix} \in \mathcal{K}$$

if and only if $\mu^2 = \sqrt{-c/\overline{b}}$. Since $bc < 0$, we have $-c/\overline{b} > 0$. Hence $\mu > 0$ is well-defined.

We still have to show that μ does not depend on the transformation $V \in G$, $V(0) \neq 0$.

Suppose that G contains the transformations

$$U = \begin{pmatrix} \varepsilon & 0 \\ 0 & \overline{\varepsilon} \end{pmatrix}, \quad |\varepsilon| = 1, \ \varepsilon \neq \pm 1,$$

and

$$V_1 = \begin{pmatrix} a_1 & b_1 \\ -\overline{b}_1 & \overline{a}_1 \end{pmatrix}, \quad |a_1|^2 + |b_1|^2 = 1, \ b_1 \neq 0.$$

Let $V_2 \in G$ such that $V_2(0) \neq 0$. Then by (A.12)

$$V_2 = \begin{pmatrix} a_2 & b_2 \\ c_2 & \overline{a}_2 \end{pmatrix}, \quad |a_2|^2 + b_2 c_2 = 1.$$

Since

$$V_1 V_2 = \begin{pmatrix} a_1 a_2 + b_1 c_2 & a_1 b_2 + \overline{a}_2 b_1 \\ \overline{a}_1 c_2 - a_2 \overline{b}_1 & \overline{a}_1 \overline{a}_2 - \overline{b}_1 b_2 \end{pmatrix},$$

we have by (A.12) $a_1 a_2 + b_1 c_2 = a_1 a_2 - b_1 \overline{b}_2$. Hence $c_2 = -\overline{b}_2$. □

Corollary A.7.7 *If $G \subset M$ is a finite group then $hGh^{-1} \subset \mathcal{K}$ for some $h \in M$.*

Proof. A finite group $G \subset M$ is purely elliptic since the powers g^n, $n = 0, \pm 1, \pm 2, \ldots$ are all distinct transformations whenever $g \neq \mathrm{id}$ is non-elliptic.
□

A.8 Discrete groups of Möbius transformations

SL(2, **C**) is a topological group with the norm $\|A\| = (|a|^2 + |b|^2 + |c|^2 + |d|^2)^{1/2}$, $A = \begin{pmatrix} a & b \\ c & d \end{pmatrix}$, $ad - bc = 1$. A subgroup $\Gamma \subset \mathrm{SL}(2,\mathbf{C})$ is *discrete*, if all subsets of Γ are open in Γ, i.e., if for every $A \in \Gamma$ there exists $\varepsilon > 0$ such that

$$\Gamma \cap \{B \in \mathrm{SL}(2,\mathbf{C}) \mid \|A - B\| < \varepsilon\} = \{A\}.$$

By Corollary A.4.2, Γ is discrete if and only if $\{I\}$ is open in Γ. Hence Γ is discrete if and only if the conditions $A_n \to I$, $A_n \in \Gamma$, imply that $A_n = I$ for all sufficiently large values of n.

$N = \{I, -I\}$ is a normal subgroup of SL(2, **C**). The quotient group PSL(2, **C**) = SL(2, **C**)/N and the group M of Möbius transformations are canonically isomorphic. If PSL(2, **C**) is equipped with the topology co–induced by the projection $P : \mathrm{SL}(2,\mathbf{C}) \to \mathrm{PSL}(2,\mathbf{C})$, then PSL(2, **C**) becomes a topological group.

A subgroup $G \subset \mathrm{PSL}(2,\mathbf{C})$ is *discrete* if and only if $\{P(I)\}$ is open in G. If $G \subset M \simeq \mathrm{PSL}(2,\mathbf{C})$ is discrete and $h \in M$, then hGh^{-1} is discrete. The norm

$$\|g\| = (|a|^2 + |b|^2 + |c|^2 + |d|^2)^{1/2}$$

of a Möbius transformation $g(z) = \frac{az+b}{cz+d}$, $ad - bc = 1$, is well–defined. Combining Theorem A.4.6 and Corollary A.4.4 we obtain

Theorem A.8.1 *A group $G \subset M$ is discrete if and only if the set $\{g \in G \mid \|g\| < t\}$ is finite for all $t > 0$.* □

Let G be a cyclic group. We may suppose that G is generated by a standard transformation m_k. Since

$$\|m_k^n\| = (|k|^n + |k|^{-n})^{1/2} \quad \text{if } k \neq 1$$

and

$$\|m_1^n\| = (2 + n^2)^{1/2},$$

it follows from Theorem A.8.1 that G is discrete if g is parabolic or loxodromic. If m_k is elliptic, i.e., $k \neq |k| = 1$, G is discrete if and only if g is of finite order.

There exist also non–cyclic Abelian discrete groups $G \subset M$:

- G is generated by translations $z \mapsto z + \omega_1$ and $z \mapsto z + \omega_2$, $\mathrm{Im}\,\frac{\omega_1}{\omega_2} \neq 0$,
- G is conjugate to the quadratic group,
- G is generated by an elliptic $h \in M$ of finite order and a loxodromic $g \in M$ such that $F_g = F_h$.

Theorem A.8.2 *Let g, $h \in M$. If g is loxodromic and $F_g \cap F_h$ contains exactly one point, then the group G generated by g and h is not discrete.*

Proof. We may suppose that $F_g \cap F_h = \{\infty\}$. Replacing g by g^{-1} if necessary we have

$$\begin{aligned} g(z) &= \alpha z, \quad |\alpha| > 1, \\ h(z) &= az + b, \quad b \neq 0. \end{aligned}$$

Then $g^{-n}(h(g^n(z))) = az + \alpha^{-n} b$ and

$$\|g^{-n} \circ h \circ g^n\| = \left(|a| + \frac{1}{|a|} + \frac{|b|^2}{|\alpha|^{2n}|a|}\right)^{1/2}.$$

Since $\|g^{-n} \circ h \circ g^n\| \to (|a| + |a|^{-1})^{1/2}$, the group G is not discrete by Theorem A.8.1. □

Theorem A.8.3 *Let $G \subset M$ be a group. If there exists an infinite sequence $(g_n) \subset G$ such that*

$$g_n(w) \to w \quad \textit{for } w = 1,\ 0,\ \infty,$$

then G is not discrete.

Proof. Let the representations

$$g_n(z) = \frac{a_n z + b_n}{c_n z + d_n}, \quad a_n d_n - b_n c_n = 1,$$

be chosen such that $\operatorname{Re} d_n \geq 0$. Since

$$g_n(1) = \frac{a_n + b_n}{c_n + d_n}, \quad g_n(0) = \frac{b_n}{d_n}, \quad g_n(\infty) = \frac{a_n}{c_n},$$

we have

$$\begin{aligned} g_n(1) - g_n(0) &= \frac{1}{c_n d_n + d_n^2}, \\ g_n(\infty) - g_n(0) &= \frac{1}{c_n d_n}, \end{aligned}$$

and hence

$$d_n^2 = \frac{1}{g_n(1) - g_n(0)} - \frac{1}{g_n(\infty) - g_n(0)} \to 1.$$

Since $\operatorname{Re} d_n \geq 0$, also $d_n \to 1$. Then

$$a_n d_n = \frac{g_n(\infty)}{g_n(\infty) - g_n(0)} \to 1$$

and it follows that $a_n \to 1$.

On the other hand,

$$c_n = \frac{a_n}{g_n(\infty)} \to 0$$

and

$$b_n = d_n g_n(0) \to 0.$$

Since the sequence (g_n) contains infinitely many distinct elements, the group G is not discrete by Theorem A.8.1. □

Let $G \subset M$ be a group. A disk D is *G–invariant* if $g(D) = D$ for all $g \in G$. If D is G–invariant, the group G is said to *act* in D. Denote by U the upper half–plane $\{z \mid \operatorname{Im} z > 0\}$.

Lemma A.8.4 *Let*

$$g(z) = \frac{az+b}{cz+d}, \qquad ad - bc = 1.$$

Then $g(U) = U$ if and only if a, b, c and d are real.

Proof. If a, b, c, d are real, g maps the extended real axis onto itself. Then $g(U)$ is either U or the lower half–plane. Since $\operatorname{Im} g(i) = 1/(c^2 + d^2)$, we have $g(U) = U$.

Conversely suppose that $g(U) = U$. Denote

$$h(z) = -\frac{1}{g(z)} = \frac{-cz-d}{az+b}.$$

Then $h(U) = U$ and $(-c)b - (-d)a = ad - bc = 1$. If $c = 0$, then $a \neq 0$. Hence, replacing g by h if necessary, we may suppose that $c \neq 0$.

Since $g(U) = U$, g is not strictly loxodromic. Then $a + d$ is real by Theorem A.6.12. Since $g(\infty) = a/c$ and $g^{-1}(\infty) = -d/c$ are real, also $(a+d)/c$ is real. Hence c, a and d are real. Finally, it follows from $ad-bc = 1$ that b is real. □

Theorem A.8.5 *If a non–Abelian purely hyperbolic group G acts in U, then G is discrete.*

Proof. We consider G as a subgroup of $\mathrm{PSL}(2, \mathbf{C})$ and use matrix notation for the transformations of G. If $A \in G$, $A = \begin{pmatrix} a & b \\ c & d \end{pmatrix}$, $ad - bc = 1$, then a, b, c, and d are real by Lemma A.8.4.

We may normalize by conjugation such that G contains a transformation A with 0 and ∞ as fixed points. Then

$$A = \begin{pmatrix} \lambda & 0 \\ 0 & \lambda^{-1} \end{pmatrix}, \quad \lambda > 0, \ \lambda \neq 0$$

by Theorem A.6.11.

Suppose that G is not discrete. Then by Corollary A.4.4 we can find a sequence $V_n \in G$ of distinct elements such that $V_n \to I$. We show that there exists an integer n_0 such that

$$V_n = \begin{pmatrix} \rho_n & 0 \\ 0 & \rho_n^{-1} \end{pmatrix}, \quad \rho_n^2 \neq 1,$$

for all $n > n_0$. To that end, let

$$\begin{aligned} C_n &= AV_nA^{-1}V_n^{-1} = \begin{pmatrix} 1 - b_nc_n(\lambda^2 - 1) & a_nb_n(\lambda^2 - 1) \\ c_nd_n(\lambda^{-2} - 1) & 1 - b_nc_n(\lambda^{-2} - 1) \end{pmatrix}, \\ D_n &= AC_nA^{-1}C_n^{-1} \end{aligned}$$

with $V_n = \begin{pmatrix} a_n & b_n \\ c_n & d_n \end{pmatrix}$, $a_nd_n - b_nc_n = 1$. Then (cf. Theorem A.7.2)

$$\begin{aligned} \operatorname{tr} C_n &= 2 - b_nc_n(\lambda - \lambda^{-1})^2 \\ \operatorname{tr} D_n &= 2 + a_nb_nc_nd_n(\lambda - \lambda^{-1})^4. \end{aligned}$$

Since $\mathrm{SL}(2, \mathbf{C})$ is a topological group and $V_n \to I$, we have $C_n \to I$. Hence $\operatorname{tr} C_n \to 2$ and $b_nc_n \to 0$. Then $a_nd_n = 1 + b_nc_n \to 1$ and $a_nd_n > 0$ for all sufficiently large values of n.

Since G is purely hyperbolic, we have $\operatorname{tr}^2 C_n \geq 4$ and $\operatorname{tr}^2 D_n \geq 4$. Since $b_nc_n(\lambda - \lambda^{-1})^2 \to 0$, there exists an n_0' such that

$$b_nc_n \leq 0 \quad \text{for } n > n_0'.$$

Similarly $a_nb_nc_nd_n \geq 0$ for sufficiently large values of n. Since $a_nd_n > 0$ for large values of n, we can find an n_0'' such that

$$b_nc_n \geq 0 \quad \text{for } n > n_0''.$$

Hence

$$b_nc_n = 0 \quad \text{for } n > n_0 = \max(n_0', n_0'').$$

Moreover, $\operatorname{tr} C_n = 2$ for $n > n_0$. Since G contains no parabolic elements, we have $C_n = I$ for $n > n_0$. Since G contains no elliptic elements, we have by Theorem A.7.3, $F_A = F_{V_n}$ and $b_n = c_n = 0$ for $n > n_0$.

Since G is non–Abelian, G contains an element B for which $F_B \cap F_A = \emptyset$ (Theorem A.7.4). Hence

$$B = \begin{pmatrix} \alpha & \beta \\ \gamma & \delta \end{pmatrix}, \quad \alpha\delta - \beta\gamma = 1, \ |\alpha + \delta| > 2, \ \beta \neq 0, \ \gamma \neq 0.$$

If

$$X_n = V_n B V_n^{-1} B^{-1} = \begin{pmatrix} \bullet & (\rho_n^2 - 1)\alpha\beta \\ (\rho_n^{-2} - 1)\gamma\delta & \bullet \end{pmatrix},$$

then $X_n \to I$. Then it follows similarly as above that

$$X_n = \begin{pmatrix} \bullet & 0 \\ 0 & \bullet \end{pmatrix}$$

for sufficiently large values of n. Since $\rho_n^2 \neq 1 \neq \rho_n^{-2}$, we have $\alpha\beta = \gamma\delta = 0$. Suppose that $\alpha = 0$. From $\alpha\delta - \beta\gamma = 1$ it follows that $\gamma \neq 0$. Hence $\delta = 0$ which contradicts the condition $|\alpha + \delta| > 2$. If $\alpha \neq 0$, then $\beta = 0$ which is also impossible. Hence there exists no sequence $V_n \to I$ and G is discrete. □

Theorem A.8.5 can be complemented as follows:

Theorem A.8.6 *If a group $G \subset M$ is purely hyperbolic, there exists a G–invariant disk D.*

Proof. Let $g, h \in G \setminus \{\mathrm{id}\}$. By Theorem A.7.2, either $F_g = F_h$ or $F_g \cap F_h = \emptyset$.

If $F_g = F_h$ for all g and h, any disk D whose boundary contains F_g is G–invariant.

Suppose that there exist $g, h \in G \setminus \{\mathrm{id}\}$ such that $F_g \cap F_h = \emptyset$. We may suppose that $F_g = \{0, \infty\}$. Let $f \in G$ and

$$g = \begin{pmatrix} u & 0 \\ 0 & 1/u \end{pmatrix}, \quad u > 0, \ u \neq 1,$$

$$f = \begin{pmatrix} \alpha & \beta \\ \gamma & \delta \end{pmatrix}, \quad \alpha\delta - \beta\gamma = 1.$$

Since f and $g \circ f$ are hyperbolic, the traces

$$t_1 = \operatorname{tr} f = \alpha + \delta$$

and

$$t_2 = \operatorname{tr}(g \circ f) = \alpha u + \delta/u$$

are real by Theorem A.6.12. Then it follows that α and δ are real.

Let

$$h = \begin{pmatrix} a & b \\ c & d \end{pmatrix}, \qquad ad - bc = 1.$$

Then a and d are real and $(a+d)^2 > 4$. Consider the fixed points x and y of h:

$$\begin{aligned} x &= \frac{a - d + \sqrt{(a+d)^2 - 4}}{2c} \\ y &= \frac{a - d - \sqrt{(a+d)^2 - 4}}{2c}. \end{aligned}$$

Since $x \neq \infty \neq y$, we have $c \neq 0$, and

$$\frac{x}{y} = \frac{a - d + \sqrt{(a+d)^2 - 4}}{a - d - \sqrt{(a+d)^2 - 4}}$$

is real. It follows that the fixed points of g and h lie on the same line through the origin. If we conjugate by a rotation $z \mapsto e^{i\vartheta}z$, the matrix of g is not changed but the fixed points x and y of h can be mapped e.g. on the real axis. Hence we may suppose that g and h map U onto itself. Then by Lemma A.8.4, a, b, c and d are real.

If $f \in G$, $f = \begin{pmatrix} \alpha & \beta \\ \gamma & \delta \end{pmatrix}$, $\alpha\delta - \beta\gamma = 1$, then α and δ are real and

$$f \circ h = \begin{pmatrix} \alpha a + \beta c & \bullet \\ \bullet & \gamma b + \delta d \end{pmatrix}.$$

Since $f \circ h \in G$, also $\alpha a + \beta c$ and $\gamma b + \delta d$ are real. Since $bc \neq 0$, it follows that also β and γ are real. then $f(U) = U$ by Lemma A.8.4. □

Combining Theorems A.8.5 and A.8.6 we obtain a quite useful result on purely hyperbolic groups.

Theorem A.8.7 *A non–Abelian purely hyperbolic group is discrete.* □

Appendix B

Traces of matrices

B.1 Trace functions

In order to complete certain arguments of Section 4.12 we need to consider traces of matrices. Here we review the results of Heinz Helling ([39]).

Definition B.1.1 *Let Γ be a group. A function $t : \Gamma \to \mathbf{R}$, not identically 0, satisfying*

$$t(\alpha\beta) + t(\alpha^{-1}\beta) = t(\alpha)t(\beta) \tag{B.1}$$

for any $\alpha, \beta \in \Gamma$, is called a trace function.

Let $\theta : \Gamma \to \mathrm{SL}_2(\mathbf{R})$ be an injective homomorphisms. A computation shows that $t = \operatorname{tr}\theta$ is a trace function on Γ. Here $\operatorname{tr}\theta$ is the usual trace of a matrix.

It is a straightforward verification to show that a trace function has the following properties:

1. Let ε be the identity element of the group Γ. Then $t(\varepsilon) = 2$.

2. $t(\alpha^{-1}) = t(\alpha)$.

3. $t(\alpha\beta) = t(\beta\alpha)$.

4. Permuting the arguments of the function

   $$\tilde{t}(\alpha, \beta, \gamma) = 2t(\alpha\beta\gamma) - t(\alpha)t(\beta\gamma) - t(\beta)t(\gamma\alpha) - t(\gamma)t(\alpha\beta) + t(\alpha)t(\beta)t(\gamma)$$

   its value gets multiplied by the sign of the permutation.

5. For $\alpha, \beta \in \Gamma$, $n \in \mathbf{Z}$, $t(\alpha^n\beta)$ is a polynomial in $t(\alpha)$, $t(\beta)$, and $t(\alpha\beta)$ with rational coefficients.

6. Let w be a word in $\alpha_1, \ldots, \alpha_n \in \Gamma$ and t a trace function. Then $t(w)$ is a polynomial of the values

$$t(\alpha_{j_1} \cdot \alpha_{j_2} \cdot \ldots \cdot \alpha_{j_i}),\ 1 \leq j_1 < j_2 < \cdots < j_i \leq n,\ 1 \leq i \leq n.$$

7. Let α, β, γ, $\delta \in \Gamma$. The value of a trace function at $\alpha\beta\gamma\delta$ satisfies

$$\begin{aligned} 2t(\alpha\beta\gamma\delta) &= t(\alpha)t(\beta\gamma\delta) + t(\beta)t(\alpha\gamma\delta) + t(\gamma)t(\alpha\beta\delta) + t(\delta)t(\alpha\beta\gamma) \\ &\quad + t(\alpha\beta)t(\gamma\delta) - t(\alpha\gamma)t(\beta\delta) + t(\alpha\delta)t(\beta\gamma) \\ &\quad - t(\alpha)t(\beta)t(\gamma\delta) - t(\gamma)t(\delta)t(\alpha\beta) \\ &\quad - t(\alpha)t(\delta)t(\beta\gamma) - t(\beta)t(\gamma)t(\alpha\delta) \\ &\quad + t(\alpha)t(\beta)t(\gamma)t(\delta). \end{aligned}$$

In particular, $t(\alpha\beta\gamma\delta)$ is a polynomial with rational coefficients of the values of the trace function t at products of at most three elements of $\{\alpha, \beta, \gamma, \delta\}$.

8. Let α, $\beta \in \Gamma$. For a trace function t, let $k_t(\alpha, \beta) = abc - a^2 - b^2 - c^2 + 4$ where $a = t(\alpha)$, $b = t(\beta)$, $c = t(\alpha\beta)$. Let

$$\kappa(\alpha, \beta; u, v) = \alpha^u \beta^v \alpha^{-u} \beta^{-v},\ u, v = \pm 1,$$

be a commutator of α and β. Then

$$t(\kappa(\alpha, \beta; u, v)) = t(\kappa(\beta, \alpha; u, v)) = 2 - k_t(\alpha, \beta).$$

Let F be a topological group. We use the notation $\mathrm{Aut}(F)$ for the group of continuous automorphisms of F.

Theorem B.1.1 ([39, Proposition 1]) *Let Γ be a group, $t : \Gamma \to \mathbf{R}$ a trace function such that $k_t : \Gamma \times \Gamma \to \mathbf{R}$ is not identically 0. We suppose, furthermore, that $k_t(\alpha, \beta) \leq 0$ if $|t(\alpha)| \leq 2$. There exists a representation $\theta : \Gamma \to \mathrm{SL}_2(\mathbf{R})$ such that $t = \mathrm{tr}\,\theta$. If θ_1 and θ_2 are both such representations, then there exists an $g \in \mathrm{Aut}(\mathrm{SL}_2(\mathbf{R}))$ such that $\theta_2 = g \circ \theta_1$.*

Proof. Let α, $\beta \in \Gamma$ be elements for which $k_t(\alpha, \beta) \neq 0$. Let A, $B \in \mathrm{SL}_2(\mathbf{R})$ be matrices for which

$$t(\alpha) = \mathrm{tr}\, A, \quad t(\beta) = \mathrm{tr}\, B, \quad t(\alpha\beta) = \mathrm{tr} AB.$$

The condition $k_t(\alpha, \beta) \neq 0$ guarantees that a matrix C is uniquely defined by the numbers $\mathrm{tr}\, C$, $\mathrm{tr}\, AC$, $\mathrm{tr}\, BC$, $\mathrm{tr}\, ABC$.

For every $\gamma \in \Gamma$ we form the equations

$$\begin{aligned} \operatorname{tr}\, \theta(\gamma) &= t(\gamma) \\ \operatorname{tr}\, A\theta(\gamma) &= t(\alpha\gamma) \\ \operatorname{tr}\, B\theta(\gamma) &= t(\beta\gamma) \\ \operatorname{tr}\, AB\theta(\gamma) &= t(\alpha\beta\gamma). \end{aligned}$$

The condition $k_t(\alpha, \beta) \neq 0$ implies that the above system of equations has a unique solution $\theta(\gamma) \in \mathrm{SL}_2(\mathbf{R})$. It is a straightforward verification to check that $\Gamma \to \mathrm{SL}_2(\mathbf{R})$, $\gamma \mapsto \theta(\gamma)$, is a homomorphism.□

Let $t : \Gamma \to \mathbf{R}$ be a function. The *topology induced by t on G* is the weakest topology of G for which G is a topological group and $t : G \to \mathbf{R}$ continuous.

Corollary B.1.2 *Assume that Γ is not commutative, t is a trace function on Γ satisfying $k_t(\alpha, \beta) \leq 0$ whenever $|t(\alpha)| \leq 2$. If the topology induced by t on Γ is Hausdorff, then there exists a faithful representation $\theta : \Gamma \to \mathrm{SL}_2(\mathbf{R})$ with $t = \operatorname{tr}\theta$. If the induced topology is discrete and the virtual cohomological dimension of Γ is 2, then $\mathrm{SL}_2(\mathbf{R})/\theta(\Gamma)$ is compact.*

One has to find α and $\beta \in \Gamma$ such that $k_t(\alpha, \beta) \neq 0$. Assume that α and β do not commute. Since the topology induced by t on Γ is Hausdorff, there exists an element $\gamma \in \Gamma$ such that $t(\alpha\beta\alpha^{-1}\beta^{-1}\gamma) \neq t(\gamma)$. If this holds already for $\gamma = \varepsilon$, then $k_t(\alpha, \beta) = 2 - t(\alpha\beta\alpha^{-1}\beta^{-1}) \neq 0$. If $t(\alpha\beta\alpha^{-1}\beta^{-1}) = 2$, then a computation shows that $k_t(\alpha\beta\alpha^{-1}\beta^{-1}, \gamma) \neq 0$. Therefore we can apply the preceding theorem to find a representation $\theta : G \to \mathrm{SL}_2(\mathbf{R})$ such that $t(\gamma) = \operatorname{tr}\theta(\gamma)$ for all $\gamma \in \Gamma$. This homomorphism is injective because of the assumption concerning the topology induced by t on Γ.

The last assertion about the compactness of $\mathrm{SL}_2(\mathbf{R})/\theta(\Gamma)$ is an immediate application of the results in [83, Corollary to Proposition 18].

Bibliography

[1] William Abikoff. *The Real Analytic Theory of Teichmüller Space.* Number 820 in Lecture Notes in Mathematics. Springer–Verlag, Berlin–Heidelberg–New York, 1980.

[2] Lars Ahlfors and Lipman Bers. Riemann's mapping theorem for variable metrics. *Ann. Math.*, 72:385 – 404, 1960.

[3] Lars Ahlfors and Leo Sario. *Riemann Surfaces.* Princeton University Press, Princeton, New Jersey, 1960.

[4] Lars V. Ahlfors. On quasiconformal mappings. *J. Analyse Math.*, III:1 – 58, 1954.

[5] Lars V. Ahlfors. The complex analytic structure of the space of closed Riemann surfaces. In Rolf Nevanlinna et. al., editor, *Analytic Functions*, pages 45 – 66. Princeton University Press, 1960.

[6] Lars V. Ahlfors. *Lectures on quasiconformal mappings.* The Wadsworth & Brooks / Cole Mathematics Series. Wadsworth & Brooks / Cole Advanced Books & Software, Monterey, California, 1987. Manuscr. prep. with the assist. of Clifford J. Earle jun. (Reprint).

[7] Norman L. Alling. *Real Elliptic Curves.* Number 54 in Mathematics Studies, Notas de Matemática. North–Holland, Amsterdam • New York • Oxford, 1981.

[8] Norman L. Alling and Newcomb Greenleaf. *Foundations of the Theory of Klein Surfaces*, volume 219 of *Lecture Notes in Mathematics.* Springer–Verlag, Berlin–Heidelberg–New York, 1971.

[9] Aldo Andreotti and Per Holm. Quasianalytic and parametric spaces. In Per Holm, editor, *Real and complex singularities, Oslo 1976*, pages 13 – 97, Alphen aan den Rijn, The Netherlands, 1977. Sijthoff & Noordhoff International Publishers.

[10] Alan F. Beardon. *The Geometry of Discrete Groups.* Number 91 in Graduate Texts in Mathematics. Springer–Verlag, Berlin–Heidelberg–New York, 1983.

[11] Lipman Bers. Spaces of degenerating Riemann surfaces. In Leon Greenberg, editor, *Discontinous Groups and Riemann Surfaces, Proc. 1973 Conf. Univ. Maryland*, number 79 in Ann. Math. Stud., pages 43 – 55, Princeton, NJ, 1974. Princeton University Press and University of Tokyo Press.

[12] Lipman Bers. Finite dimensional Teichmüller spaces and generalizations. *Bull. Amer. Math. Soc.*, 5(2):131 – 172, September 1981.

[13] Lipman Bers. An Inequality for Riemann Surfaces. In Isaac Chavel and Hersel M. Farkas, editors, *Differential Geometry and Complex Analysis*, pages 87 – 93. Springer–Verlag, Berlin–Heidelberg–New York, 1985.

[14] N. Bourbaki. *Variétés différentielles et analytiques.* Eléments de mathématique XXXIII. Hermann, 1967.

[15] Robert Brooks. Constructing isospectral manifolds. *Am. Math. Mon.*, 95(9):823 – 839, 1988.

[16] G. W. Brumfiel. Quotient spaces for semialgebraic equivalence relations. Manuscript, to appear in Math. Z.

[17] Emilio Bujalance, José J. Etayo, José M. Gamboa, and Grzegorz Gromadzki. *Automorphism Groups of Compact Bordered Klein surfaces, A Combinatorial Approach.* Number 1439 in Lecture Notes in Mathematics. Springer–Verlag, Berlin–Heidelberg–New York, 1990.

[18] W. Burnside. On a Class of Automorphic Functions. *Proc. London Math. Soc.*, XXIII:49 – 88, 1891.

[19] P. Buser and K.-D. Semmler. The geometry and spectrum of the one holed torus. *Comment. Math. Helv.*, 63(2):259 – 274, 1988.

[20] Peter Buser. Sur le spectre de longueurs des surfaces de Riemann. *C.R. Acad. Sci., Paris, Ser. I*, 292:487 – 489, 1981.

[21] Peter Buser. Isospectral Riemann surfaces. *Ann. Inst. Fourier*, 36(2):167 – 192, 1986.

[22] Peter Buser. *Geometry and Spectra of Compact Riemann Surfaces.* Birkhäuser Verlag, Basel–Boston–New York, 1991. To appear.

[23] Peter Buser and Mika Seppälä. Symmetric pants decompositions of Riemann surfaces, 1991. Preprint, to appear in Duke Math. J.

[24] Henri Cartan. Exposé 12. Quotient d'un variété analytique par un groupe discret d'automorphismes. In *Seminaire Henri Cartan 1953/1954*. W. A. Benjamin, Inc., New York – Amsterdam, 1967. Séminaire E. N. S., 1953 – 1954.

[25] H.-D. Coldeway. Kanonische polygone endlich erzeugter fuchsscher gruppen, 1971. Dissertation.

[26] C. J. Earle. On moduli of closed Riemann surfaces with symmetries. In *Advances in the Theory of Riemann Surfaces. Annals of Mathematics Studies 66*, pages 119 – 130, Princeton, New Yersey, 1971. Princeton University Press and University of Tokyo Press.

[27] C. J. Earle and A. Marden. Manuscript.

[28] D. B. A. Epstein. Curves on 2–manifolds and isotopies. *Acta Math.*, 115:83 – 107, 1966.

[29] Hershel M. Farkas and Irwin Kra. *Riemann Surfaces*, volume 71 of *Graduate Texts in Mathematics*. Springer–Verlag, Berlin–Heidelberg–New York, 1980.

[30] A Fathi and F. Laudenbach, editors. *Travaux de Thurston sur les surfaces*, volume 66 – 67 of *Astérisque*. Soc. Math. France, Paris, 1979.

[31] R. Fricke and F. Klein. *Vorlesungen über die Theorie der automorphen Funktionen*. Druck und Verlag von B. G. Teubner, Leipzig, 1897, 1926.

[32] Frederick P. Gardiner. *Teichmüller Theory and Quadratic Differentials*. A Wiley–Interscience of Series of Texts, Monographs, and Tracts. John Wiley & Sons, New York Chichester Brisbane Toronto Singapore, 1987.

[33] F. W. Gehring and O. Lehto. On the total differentiability of functions of a complex variable. *Ann. Acad. Sci. Fenn. A I*, 272, 1959.

[34] Samuel I. Goldberg. *Curvature and Homology*. Dover Publications, Inc., New York, 1962.

[35] P. Gordan. Über endliche Gruppen linearer Transformationen einer Veränderlichen. *Math. Ann.*, 12:23 – 46, 1877.

[36] Noami Halpern. Some contributions to the theory of Riemann surfaces, 1978. Thesis, Columbia University.

[37] Joe Harris. An introduction to the moduli space of curves. In *Mathematical aspects of string theory*, number 1 in Adv. Ser. Math. Phys., pages 285 – 312, 1987. Proc. Conf., San Diego Calif. 1986.

[38] Joe Harris and David Mumford. On the Kodaira Dimension of the Moduli Space of Curves. *Invent. math*, 67:23 – 86, 1982.

[39] Heinz Helling. Diskrete untergruppen von $SL_2(\mathbf{R})$. *Inventiones math.*, (17):217–229, 1972.

[40] Heinz Helling. Über den Raum der kompakten Riemannschen flächen vom Geschlecht 2. *J. reine angew. Math.*, 268/269:286 – 293, 1974.

[41] Frank Herrlich. The extended Teichmüller space, 1991. Math. Z. (to appear) (Math. Inst., Ruhr-Univ., D-4630 Bochum).

[42] Adolf Hurwitz. Über Riemann'sche Flächen mit gegebenen Verzweigungspunkten. *Math. Ann.*, 39:1 – 61, 1891. Reprinted in Adolf Hurwitz: Mathematische Werke, Band I, Funktionentheorie, Birkhäuser, 1932, 321 – 381.

[43] Adolf Hurwitz. Algebraische Gebilde mit eindeutigen Transformationen in sich. *Math. Ann.*, 41:403 – 442, 1893. Reprinted in Adolf Hurwitz: Mathematische Werke, Band I, Funktionentheorie, Birkhäuser, 1932, 391 – 430.

[44] Linda Keen. Collars on Riemann surfaces. In *Discontinuous Groups and Riemann Surfaces*, volume 79 of *Ann. of Math. Studies*, pages 263 – 268. Princeton University Press, 1974.

[45] Felix Klein. Über eine Art von Riemannschen Flächen. *Math. Annalen*, 10, 1876.

[46] Felix Klein. Über eine neue Art von Riemannschen Flächen. *Math. Annalen*, 10, 1876.

[47] Felix Klein. *Über Riemanns Theorie der algebraischen Funktionen und ihrer Integrale. — Eine Ergänzung der gewöhnlichen Darstellungen.* B. G. Teubner, Leipzig, 1882.

[48] Felix Klein. Über Realitätsverhältnisse bei der einem beliebigen Geschlechte zugehörigen Normalkurve der φ. *Math. Annalen*, 42, 1892.

[49] Felix Klein. Über Realitätsverhältnisse bei der einem beliebigen Geschlechte zugehörigen Normalkurve der φ. *Math. Annalen*, 42, 1892.

[50] Finn Knudsen and David Mumford. The projectivity of the moduli space of stable curves. I: Preliminaries on "det" and "Div". *Math. Scand.*, 39:19 – 55, 1976.

[51] Finn F. Knudsen. The projectivity of the moduli space of stable curves. II: The stacks $M\ sub(g,n)$. *Math. Scand.*, 52:161 – 199, 1983.

[52] Finn F. Knudsen. The projectivity of the moduli space of stable curves. III: The line bundles on $M\ sub(g,n)$, and a proof of the projectivity of $M - sub(g,n)$ in characteristic 0. *Math. Scand.*, 52:200 – 212, 1983.

[53] Irwin Kra. On Lifting of Kleinian Groups to $SL(2,\mathbf{C})$. In Isaac Chavel and Hersel M. Farkas, editors, *Differential Geometry and Complex Analysis*, pages 181 – 193. Springer–Verlag, 1985.

[54] S. Kravetz. On the geometry of Teichmüller spaces and the structure of their modular groups. *Ann. Acad. Sci. Fenn.*, 278:1–35, 1959.

[55] Ravi S. Kulkarni. Some investigations on symmetries of Riemann surfaces. *Institut Mittag-Leffler Report*, (8), 1989/90.

[56] A. G. Kurosh. *The Theory of Groups, 1–2.* Chelsea Publishing Company, New York, 1955–1956.

[57] Joseph Lehner. *Discontinuous Groups and Automorphic Functions.* Mathematical Surveys. American Mathematical Society, Providence, Rhode Island, 1964.

[58] O. Lehto. Quasiconformal homeomorphisms and Beltrami equations. In W. J. Harvey, editor, *Discrete Groups and Automorphic Functions*, pages 121 – 142. Academic Press, 1977.

[59] O. Lehto and K. I. Virtanen. *Quasiconformal mappings in the plane*, volume 126 of *Die Grundlehren der mathematischen Wissenschaften.* Springer–Verlag, Berlin–Heidelberg–New York, 1973. Translated from the German by K. W. Lucas. 2nd ed.

[60] Olli Lehto. *Univalent Functions and Teichmüller Spaces*, volume 109 of *Graduate Texts in Mathematics.* Springer-Verlag, Berlin–Heidelberg–New York, 1986.

[61] W. Magnus, A. Karras, and D. Solitar. *Combinatorial group theory: presentations of groups in terms of generators and relations.* Interscience, 1966.

[62] H. P. McKean. Selberg's trace formula as applied to a compact Riemann surface. *Commun. pure appl. Math.*, 25:225 – 246, 1972.

[63] H. P. McKean. Correction to: Selberg's trace formula as applied to a compact Riemann surface. *Commun. pure appl. Math.*, 27:134, 1974.

[64] J. Morgan and P. Shalen. Valuations, trees, and degenerations of hyperbolic structures. I. *Ann. of Math*, pages 401–476, 1984.

[65] D. Mumford. The structure of the moduli spaces of curves and abelian varieties. In *Proc. Internat. Congr. Math.*, pages 457 – 467, 1970.

[66] D. Mumford and J. Fogarty. *Geometric invariant theory*, volume 34 of *Ergebnisse der Mathematik.* Springer–Verlag, Berlin–Heidelberg–New York, 1982.

[67] David Mumford. *Curves and Their Jacobians.* University of Michigan Press, Ann Arbor, Michigan, second printing 1976 edition, 1975.

[68] David Mumford. Stability of projective varieties. *L'Ens. Math.*, 23:39 – 110, 1977.

[69] Subhashis Nag. *The complex analytic theory of Teichmueller spaces.* Canadian Mathematical Society Series of Monographs and Advanced Texts. John Wiley, New York etc., 1988.

[70] S. M. Natanzon. Moduli spaces of real curves. *Trans. Mosc. Math. Soc.*, 1:233 – 272, 1980.

[71] H. L. Royden. Automorphisms and isometries of Teichmüller space. In Cabiria Andreian Cazacu, editor, *Proceedings of the Romanian-Finnish Seminar on Teichmüller Spaces and Quasiconformal Mappings, Brasov 1969*, pages 273 – 286. Publishing House of the Academy of the Socialist Republic of Romania, 1971.

[72] H.L. Royden. Automorphisms and isometries of Teichmüller space. In Lars V. Ahlfors et al., editor, *Advances in the Theory of Riemann Surfaces*, volume 66 of *Ann. of Math. Studies*, pages 369–383. 1971.

[73] Niels Schwartz. Compactification of varieties. *Arkiv för matematik*, 28(2):333 – 370, 1990.

[74] Mika Seppälä. Teichmüller Spaces of Klein Surfaces. *Ann. Acad. Sci. Fenn. Ser. A I Mathematica Dissertationes*, 15:1 – 37, 1978.

[75] Mika Seppälä. Quotients of complex manifolds and moduli spaces of Klein surfaces. *Ann. Acad. Sci. Fenn. Ser. A. I. Math.*, 6:113 – 124, 1981.

[76] Mika Seppälä. On moduli of real curves. In Jean-Jacques Risler, editor, *Seminaire sur la Geometrie Algebrique Reelle*, pages 85 – 95, Paris VII, 1986. Publications Mathématiques de l'Université Paris VII.

[77] Mika Seppälä. Moduli spaces of stable real algebraic curves. *Institut Mittag-Leffler Report*, 12:1 – 28, 1989/90. To appear in: Ann. scient. Éc. Norm. Sup. 4e série.

[78] Mika Seppälä. Real algebraic curves in the moduli space of of complex curves. *Compositio Mathematica*, 74:259 – 283, 1990.

[79] Mika Seppälä and Robert Silhol. Moduli spaces for real algebraic curves and real abelian varieties. *Math. Z.*, 201:151 – 165, 1989.

[80] Mika Seppälä and Tuomas Sorvali. Parametrization of Teichmüller spaces by geodesic length functions. In D. Drasin, C. J. Earl, F. W. Gehring, I. Kra, and A. Marden, editors, *Holomorphic Functions and Moduli II*, volume 11 of *Publications of the Mathematical Sciences Research Institute Berkeley*, pages 267 – 283. Springer–Verlag, New York Berlin Heidelberg London Paris Tokyo, 1988.

[81] Mika Seppälä and Tuomas Sorvali. Horocycles on Riemann surfaces, May 1991. Preprint, to appear in Proc. Amer. Math. Soc.

[82] Mika Seppälä and Tuomas Sorvali. Traces of commutators of Möbius transformations, 1991. To appear in Math. Scand.

[83] J. P. Serre. Cohomologie des groupes discrets. In *Prospects of mathematics*, volume 70, pages 77 – 169. 1971.

[84] G. Shimura. On the fields of rationality for an abelian variety. *Nagoya Math. J.*, 45:167 – 178, 1972.

[85] C. L. Siegel. Bemerkung zu einem Satz von Jakob Nielsen. In K. Chandrasekharan und H. Maaß, editor, *Carl Ludwig Siegel Gesammelte Abhandlungen, Band III*, pages 92 – 96. Springer–Verlag, Berlin Heidelberg New York, 1966.

[86] Carl L. Siegel. *Topics in Complex Function Theory*, volume 2 of *Wiley-Interscience*. John Wiley & Sons, Inc., New York - London - Sydney - Toronto, 1971.

[87] Carl L. Siegel. *Topics in Complex Function Theory*, volume 1 of *Wiley-Interscience*. John Wiley & Sons, Inc., New York - London - Sydney - Toronto, 1971.

[88] Robert Silhol. Compactifications of Moduli Spaces in Real Algebraic Geometry, 1990. Preprint, to appear in Inv. Math.

[89] Tuomas Sorvali. The boundary mapping induced by an isomorphism of covering groups. *Ann. Acad. Sci. Fenn.*, 526:1 - 31, 1972.

[90] Tuomas Sorvali. On discontinuity of Moebius groups without elliptic elements. *Publ. Univ. Joensuu, Ser. B 9*, pages 1 - 4, 1974.

[91] Edwin H. Spanier. *Algebraic Topology*. McGraw-Hill Book Company, New York, 1966.

[92] Kurt Strebel. On the maximal dilatation of quasiconformal mappings. *Proc. Amer. Math. Soc.*, 6:903 - 909, 1955.

[93] Oswald Teichmüller. Extremale quasikonforme Abbildungen und quadratische Differentiale. In Lars V. Ahlfors and Frederick W. Gehring, editors, *Oswald Teichmüller Gesammelte Abhandlungen Collected Papers*, pages 335 - 531. Springer-Verlag, Berlin-Heidelberg-New York, 1982. Originally appeared in Abh. Preuß. Akad. Wiss., math.-naturw. Kl. 22, 197(1939).

[94] Oswald Teichmüller. Veränderliche Riemannsche Flächen. In Lars V. Ahlfors and Frederick W. Gehring, editors, *Oswald Teichmüller Gesammelte Abhandlungen Collected Papers*, pages 712 - 727. Springer-Verlag, Berlin-Heidelberg-New York, 1982. Originally appeared in Deutsche Math. 7, 344 - 359 (1944).

[95] Marie-France Vigneras. Varietes riemanniennes isospectrales et non isometriques. *Ann. Math.*, (112):21 - 32, 1980.

[96] Guido Weichhold. Über symmetrische Riemannsche Flächen und die Periodizitätsmodulen der zugehörigen Abelschen Normalintergale erstes Gattung. *Leipziger Dissertation*, 1883.

[97] A. Weil. On discrete subgroups of Lie groups I. *Ann. Math.*, 72:369 - 384, 1960.

[98] A. Weil. On discrete subgroups of Lie groups II. *Ann. Math.*, 75:578 – 602, 1962.

[99] A. Wiman. Über die algebraischen Kurven von den Geschlechten $p = 4, 5$ und 6, welche eindeutige Transformationen in sich besitzen. *Bihang Till Kongl. Svenska Ventenskaps–Akademiens Handlingar*, 21(3):1 – 41, 1895.

[100] A. Wiman. Über die hyperelliptischen Kurven und diejenigen vom Geschlecht $p = 3$, welche eindeutige Transformationen in sich zulassen. *Bihang Till Kongl. Svenska Ventenskaps–Akademiens Handlingar*, 21(1):1 – 23, 1895.

[101] Scott A. Wolpert. The geometry of the moduli space of Riemann surfaces. *Bull. Am. Math. Soc., New Ser.*, 11:189 – 191, 1984.

[102] Li Zhong. Nonuniqueness of Geodesics in Infinite Dimensional Teichmüller spaces. *Research Report*, (29):1 – 25, 1990.

Subject Index